NATURE'S MACHINES

NATURE'S MACHINES

the story of biomechanist

MIMI KOEHL

by Deborah Parks

Franklin Watts
A Division of Scholastic Inc.
New York • Toronto • London • Auckland • Sydney
Mexico City • New Delhi • Hong Kong
Danbury, Connecticut

Joseph Henry Press
Washington, D.C.

AUTHOR'S ACKNOWLEDGMENTS

Thank you, Mimi Koehl, for seeing the beauty of science and for sharing your stories with such passion and joy. You've made me laugh, cry, and learn about a whole new field. Thanks, too, to all the people who added their own tales—your husband Zack Powell, your brother Bob Koehl, your mentors Steve Wainwright and Bob Paine, and your sisters in science, Ellen Daniell and Suzanne McKee. I also value the support from my editor, Allan Fallow. Finally, I'm extremely grateful to three special people: Haroldine Gold Keeler, the mother who has always supported my adventures; Darrell Kozlowski, my wonderful publishing friend who advises and guides me; and Richard Parks, the husband who listens to late-night manuscript readings. —DP

Cover photo: Biomechanist Mimi Koehl stands beside an algae-covered rock on a beach in California. A camera is one of the basic research tools she uses to study the structure and movement of living things.

Cover design: Michele de la Menardiere

Library of Congress Cataloging-in-Publication Data

Parks, Deborah A., 1948-
Nature's machines : the story of biomechanist Mimi Koehl / by Deborah Amel Parks.
p. cm. — (Women's adventures in science)
Includes bibliographical references (p.) and index.
ISBN 0-531-16780-1 (lib. bdg.) ISBN 0-309-09559-X (trade pbk.) ISBN 0-531-16955-3 (classroom pbk.)
1. Koehl, Mimi, 1948—Juvenile literature. 2. Biologists—United States—Juvenile literature. 3. Women biologists—United States—Juvenile literature. 4. Biomechanics—Juvenile literature. I. Title. II. Series.

QH31.K62P37 2005
570'.92—dc22

2005010201

Printed in the United States of America.
1 2 3 4 5 6 7 8 9 10 R 14 13 12 11 10 09 08 07 06 05

About the Series

The stories in the *Women's Adventures in Science* series are about real women and the scientific careers they pursue so passionately. Some of these women knew at a very young age that they wanted to become scientists. Others realized it much later. Some of the scientists described in this series had to overcome major personal or societal obstacles on the way to establishing their careers. Others followed a simpler and more congenial path. Despite their very different backgrounds and life stories, these remarkable women all share one important belief: the work they do is important and it can make the world a better place.

Unlike many other biography series, *Women's Adventures in Science* chronicles the lives of contemporary, working scientists. Each of the women profiled in the series participated in her book's creation by sharing important details about her life, providing personal photographs to help illustrate the story, making family, friends, and colleagues available for interviews, and explaining her scientific specialty in ways that will inform and engage young readers.

This series would not have been possible without the generous assistance of Sara Lee Schupf and the National Academy of Sciences, an individual and an organization united in the belief that the pursuit of science is crucial to our understanding of how the world works and in the recognition that women must play a central role in all areas of science. They hope that *Women's Adventures in Science* will entertain and enlighten readers with stories of intellectually curious girls who became determined and innovative scientists dedicated to the quest for new knowledge. They also hope the stories will inspire young people with talent and energy to consider similar pursuits. The challenges of a scientific career are great but the rewards can be even greater.

Other Books in the Series

Beyond Jupiter: The Story of Planetary Astronomer Heidi Hammel

Bone Detective: The Story of Forensic Anthropologist Diane France

Forecast Earth: The Story of Climate Scientist Inez Fung

Gene Hunter: The Story of Neuropsychologist Nancy Wexler

Gorilla Mountain: The Story of Wildlife Biologist Amy Vedder

People Person: The Story of Sociologist Marta Tienda

Robo World: The Story of Robot Designer Cynthia Breazeal

Space Rocks: The Story of Planetary Geologist Adriana Ocampo

Strong Force: The Story of Physicist Shirley Ann Jackson

Contents

INTRODUCTION

Dynamic Nature

How do the seaweeds and creatures clinging to a rugged coastline survive the crashing waves without breaking? How do animals glide, and what makes some more maneuverable than others? How do antennae catch odors wafting in the wind or in water swirling around them? How can soft, squishy animals move around without a bony skeleton?

These are some of the questions Mimi Koehl investigates. She works in a field of science called biomechanics. Mimi uses the laws of physics to study how living things move around in their environments and how they interact with the water or air that surrounds them. She wants to understand how the shape, size, or stiffness of an organism affects how well it performs particular tasks. Mimi doesn't just work in her laboratory. She gets out in nature, nose-to-nose with the creatures she studies. That way she can measure what the physical environment is like for the living machines that she analyzes.

Today Mimi teaches and does research at the University of California at Berkeley. Mimi loves being a scientist, but she had to struggle to become one. She battled resistance from some family members who thought girls should not be scientists. She also had difficulty reading because she has dyslexia. But Mimi overcame all of these obstacles to become a world-class scientist. Here's the story of how she did it.

Mimi's *achievement*

went far beyond

what was *expected*

of her as a child.

1

HOPE CALLING

In July 1990 scientist Dr. Mimi Koehl checked a voice message on the answering machine in her office at the University of California, Berkeley. It said, "This is Ken Hope from the MacArthur Foundation. You've been chosen to receive a MacArthur Fellowship."

Yeah, right, thought Mimi. *Ken Hope can't possibly be a real name!* Of course, you "can hope" to win a MacArthur Fellowship. But the MacArthur Foundation only gives them to between 20 and 30 people a year. *This has to be a practical joke. Nobody is going to hand me several hundred thousand dollars to do my kind of science,* she thought.

Award-winning faculty members of the University of California at Berkeley *(opposite)*—including biomechanist Mimi Koehl *(third from left)*—are honored at a 1991 ceremony. Mimi had just won a MacArthur Fellowship for her creative science. She studies how organisms such as corals *(above)* interact with the water or air moving around them.

~ "*Who* Are *You?*"

"Mimi had every reason to think it was a practical joke," laughs Bob Paine, one of the mentors who shaped Mimi's career. Bob and Mimi have played gags on each other for years. Mimi once planted plastic pink flamingos on field sites where Bob does ecological research on an offshore island near Washington State. Bob got even by planting them on Mimi's lawn and in her office. But Bob didn't stop there. He sent an application from "F. L. Mingo" to

University of Washington marine biologist Bob Paine laughs upon discovering that Mimi has spoofed him again—this time by planting a pink-flamingo windshield screen in his car.

Mimi's graduate science program at Berkeley. F. L. Mingo had excellent grades—for a pink plastic bird.

Mimi's big brother, Bob Koehl, plays jokes on Mimi, too. She returns the favor. Their jokes come from their favorite childhood TV program, *The Three Stooges*. Bob asks, "Who else could teach you to say intelligent things like 'Nyuk, Nyuk, Nyuk'?"

Mimi was no stranger to practical jokes. Still, she couldn't wait to solve the mystery. She wondered, *Which one of my friends or family could have dreamed up this trick?*

She finally called the phone number that "Ken Hope" had left on her answering machine. The strange voice said the very same thing she'd heard on the recording. "You've been chosen to receive a MacArthur Fellowship."

Mimi strained to recognize the speaker. She finally asked, "Who *are* you?"

The voice replied, "You haven't heard a single word I've said, have you?"

"No," answered Mimi, "I'm trying to figure out who you are. This *is* a practical joke, isn't it?"

"No," he answered. "If you don't believe it's true, just call the MacArthur Foundation and ask for me. You'll find out it's all very real."

~ "It's Really True!"

As Mimi soon discovered, the MacArthur Foundation had indeed awarded her $260,000—with no strings attached. The foundation gives such "genius grants," as they're nicknamed, to people in all walks of life—writers, artists, dancers, civil rights workers, scientists, teachers, farmers, and more. The program doesn't reward the nominees for their past work. Instead it encourages them to be creative and to do original work in the future.

Mimi enlists the help of graduate student Suzy Worcester *(far left)* to empty seawater from her hip waders after a research trip into Elkhorn Lagoon on the California coast.

"If I had an idea that required spending money on equipment," explains Mimi, "I could just do it. The fellowship gave me tremendous freedom to study anything that I thought was important. Even if the approach I used might seem wacky to some people, I was free to try."

Mimi burned to tell someone the good news. She ran into the lab where her graduate students worked.

"I got a MacArthur Fellowship," she blurted out.

"How could you have?" her student, Suzy Worcester, asked. "You're not an ecologist."

Suzy had confused the MacArthur Fellowship with the Robert H. MacArthur Award, given every two years to an ecologist who has done outstanding work in studying the environment. But Mimi is a biomechanist, not an ecologist.

"The fellowship gave me tremendous freedom to study anything that I thought was important."

Suzy's question burst Mimi's bubble, but not for long. Local newspapers soon picked up the story. "It was thrilling to be named in articles listing all these famous people," Mimi remembers. "I kept pinching myself to make sure it was really true."

~ In the Dark

Because people can't apply for a MacArthur Fellowship, they're usually shocked when they receive one. As Mimi puts it, "The awards come out of the blue, or so it seems."

Since 1981 the MacArthur Foundation has handed out awards to some 700 people. Every year the foundation picks a committee of several hundred people to nominate creative individuals in various fields. The foundation keeps the names of committee members secret. This allows them to speak freely about nominees.

Only a handful of people connected to the awards knew that Mimi might be chosen. One of them asked Mimi's husband, Zack Powell, to get a copy of Mimi's professional résumé. To keep Mimi in the dark, Zack made up a story about sending it to a colleague in England. But Mimi beat him to the punch. Determined to save Zack the trouble, she mailed it off herself.

Zack crept around the house, trying to get another copy without tipping off Mimi. After all, she still might not be picked.

~ Groundbreaking Science

Mimi's achievement went far beyond what was expected of her as a child. Throughout her early life, Mimi recalled her mother's warning: Smart women wind up spinsters or old maids. But Mimi's mother was wrong. Many men like smart women, and Mimi married one.

Zack studies and teaches oceanography at Berkeley. He specializes in ocean currents, global climate, and plankton—the tiny plants and animals upon which many marine food chains depend. He enjoys being married to someone who

Zack Powell, Mimi's husband, mans the tiller of a boat off the coast of California.

Steve Wainwright, Mimi's Ph.D. advisor, is a leader in the new field of science known as biomechanics. Every year, Mimi makes him another signature bow tie.

can talk about physics and calculus over breakfast.

Zack also appreciates Mimi's groundbreaking research. "By the time of the MacArthur Fellowship," he says, "people had begun to recognize that her work is special. It's one thing to think about biology. It's another thing entirely to think about mechanics or engineering. Yet Mimi is putting them together—not just in nature, but also in the laboratory. It's different and it's creative."

Steve Wainwright, Mimi's mentor and advisor for her Ph.D., agrees. "I was thrilled when Mimi won a MacArthur," he declares. "She has gone so much further than me in a field that I helped pioneer in North America."

~ *"Crazy, Creative People"*

At the time Mimi won the MacArthur Fellowship, the foundation held meetings in Chicago where the various MacArthur winners gathered. People read poetry, played music, showed videos of dance performances, or talked about their research.

"Mimi still has trouble recognizing just how good she is," Zack points out. "So she held back telling other MacArthur Fellows about her ideas."

"I was terrified to speak up," recalls Mimi. "I felt so insecure—I was a nerdy scientist among all these artists, writers, dancers, and musicians. For the first couple of meetings,I didn't present anything. I just watched other people talk or perform."

The foundation allowed each MacArthur Fellow, past and present, to bring one guest. Mimi brought Zack. Everybody wore a name tag, but the tags didn't say which person had won the award.

"Zack isn't shy like me," says Mimi. "So he struck up conversations with all these interesting people. They just assumed he was the MacArthur Fellow, and that I was the spouse quietly tagging along.

Important Matters

Two kinds of matter matter to Mimi: fluids and solids. Basically, fluids flow; solids don't. The reason lies in their molecular structure. In solids the molecules are bonded to each other. In fluids—liquids and gases—the molecules can move past each other. You can pick up a solid, but just try picking up a fluid without a container. It slips through your fingers.

Solids resist being deformed when you apply force. And when you stop applying force, they bounce back to their original shape. Try this with a rubber eraser. Push on it gently. What happens? Push on it harder until it deforms, or bends. Then stop pushing. Boing! It springs back into shape.

Fluids are another matter. When you apply force to a fluid, like moving your hand through bathwater, it deforms, too. But when you stop applying the force, the fluid stays deformed. It doesn't snap back to where it was before you stirred things up.

Here's another difference: Solids care about how hard, or far, you deform them. Fluids care about how fast you push them. Try pushing your hand through water slowly, then rapidly. The faster you push, the more the water (or air) resists.

So why does all this matter to Mimi? Because she's curious about how solid organisms interact with moving fluids. She wants to figure out questions like: How do living creatures stand up to fluid forces? How do they move through fluids? How do they catch things like food or smells from the air or water around them? That's why physics matters to biomechanists like Mimi.

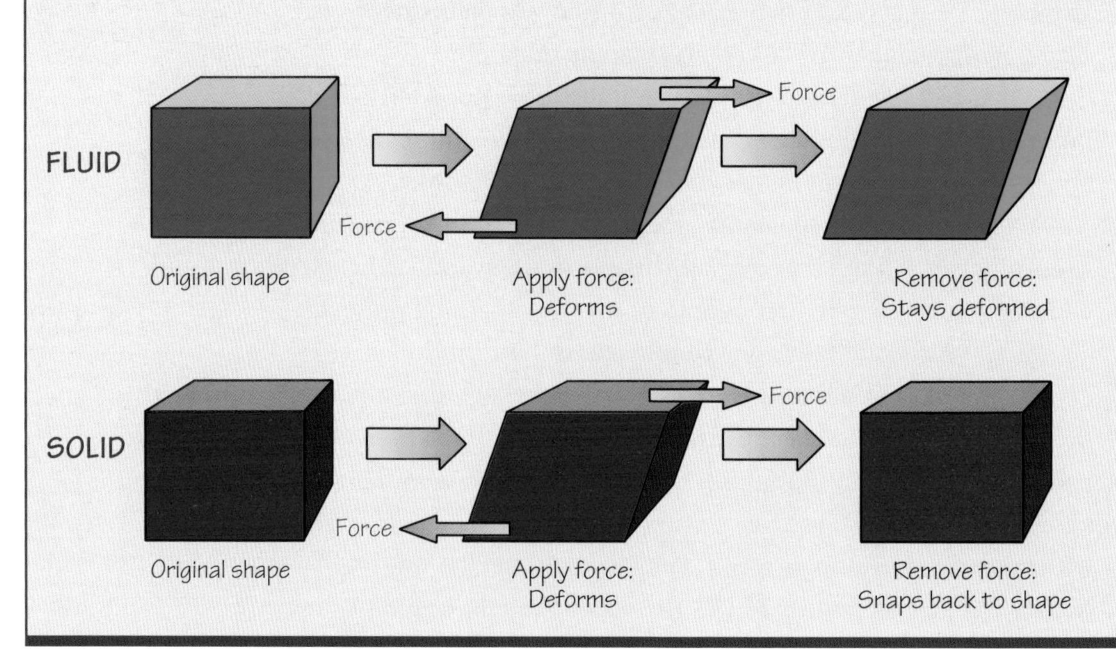

Still, it was wonderful to be surrounded by crazy, creative people doing neat stuff. I found it all very exciting."

One especially memorable person turned out to be Jack Horner, a famous scientist who was the inspiration for a character in the movie *Jurassic Park*. But it wasn't the Hollywood connection that made an impression on Mimi. It was something Jack told her. Jack was the first person to tell Mimi that she might be dyslexic. (Dyslexia is a condition that makes it extremely difficult to read, write, and spell.) Jack could recognize the signs for dyslexia because he himself is dyslexic. This was a very important observation for Mimi because it answered many questions for her. It also showed her that she was not alone.

One of America's best-known paleontologists, Jack Horner, proudly displays the skeleton of a *Tyrannosaurus rex* unearthed in Montana in 1990. Like Mimi, Jack is a MacArthur Fellow.

After meeting so many inspiring Fellows, Mimi finally summoned the courage to do a presentation on biomechanics—the exciting branch of science that uses the principles of engineering and physics to figure out how living things work.

Mimi showed videotapes of the underwater world she visits often—and loves deeply. "I didn't explain all the physics," says Mimi. "I mainly wanted people to see the shapes and forms that drew me into the ocean. I wanted them to see the beauty of water, plants, and animals in motion. Fluid flow is incredible! So that's what I did: I took them into the sea."

Diving into her work, Mimi studies spawning sea creatures on Australia's Great Barrier Reef in 1995.

Mimi's journey through life has done its own share of twisting, turning, and drifting. Some of the obstacles she faced, such as dyslexia, were so challenging that others might have quit. But Mimi's curiosity got the best of her. "I had this burning question," smiles Mimi. "How does nature work?"

Bob nicknamed his sister

"Squirt"

because she was

so much smaller—and **younger**—

than he was.

2

"GIRL STUFF"

When Mimi was a little girl her father almost had to drag her brother, Bob, into the basement workshop. "My eyes would just glaze over," Bob laughs. "I'd stand around until I could ask to leave. My sister was another matter."

Mimi couldn't wait to get into the machine-filled basement. "Carpentry fascinated me," she recalls. "My dad noticed this and showed me how to use some of his tools. Then he would give me little jobs to do, like hand-sanding or sawing. I spent hours helping him build things."

Her father also worked with Mimi on measurement and design. But, as she remembers today, "I was allowed to make only doll furniture, because that was a girl thing."

Seven-month-old Mimi *(opposite)* celebrates Easter 1949 with her father, George, and her brother, Bob. Above, she gets into the swing of things in her Maryland back-yard at age 5.

~ *What a "Squirt!"*

Mimi escaped "girl things" when she hung around with her big brother. "If my sister wanted to play with me," recalls Bob, "she had to play my kind of games. We'd build forts from toy bricks. Then we'd knock them over with tanks or a little submarine that shot wooden torpedoes."

Bob nicknamed his sister "Squirt" because she was so much smaller—and younger—than he was. Bob was eight years older than Mimi, who was born on October 1, 1948. Because Bob was a lot older, his parents called him Mimi's "junior daddy." Neither Mimi nor Bob thought much of the idea. "It was awful," they both moan.

Mimi looked up to her brother and tagged along after him whenever she could. That sometimes caused problems for Bob. "There I was, a 12-year-old kid playing with my friends," laughs Bob, "and I've got this 4-year-old squirt hanging around like my shadow."

"He'd tease me unmercifully until I went away," continues Mimi. "But I still adored him. He never said anything about the fact that I liked playing with his hand-me-down mechanical toys—trucks, airplanes, steam shovels—more than with my dolls."

Mimi and her big brother Bob *(above)* pause before the White House in Washington, D.C., in 1954.

~ *Keeping Pace with the Pack*

Mimi got away with being a tomboy while she was a little kid. She built tree houses and played softball, kickball, and dodgeball with the neighborhood boys. She also went on long hikes with her family. Her mother always said, "It's important to keep pace." Mimi, the runt of the family, churned her short legs in double time to keep up with "the big guys."

Mimi seems to have gotten a running start on life. Unlike most toddlers, as family legend has it, she skipped the walking stage entirely. Mimi got up and

Mimi sprints toward the camera *(right)* as her mother, Alma, looks on. She's already hugging a squishy creature—in this case, a teddy bear.

ran, fell down, got up and ran, and kept trying until she got it right.

She plunged into swimming with the same fearlessness. "Before I was even three," says Mimi," the family would take me to the ocean. I loved watching the waves break and feeling the water pull the sand from beneath my feet. I loved the water, and I loved swimming. I don't remember being afraid of the ocean."

Even as a 2-year-old toddler, Mimi felt at home in the sea.

Yet Mimi's parents didn't encourage her tomboy spirit as she grew older. "My parents were very traditional," Bob explains. "Even though Mimi and I had similar skills, I was programmed to be a scientist or a doctor because I was a boy. Mimi, because she was a girl, was programmed to be a nurse or a teacher. That's not a put-down of either job; it's just that my parents, especially my mother, figured Mimi should snag a man and he'd take care of her. My sister had other ideas."

~ *Portrait of an Artist*

On the surface, Mimi's parents lived the life of a typical 1950s family. Her father, George, was a physics professor and a dean at George Washington University in Washington, D.C. Her mother, Alma, stayed home and took care of the children in the Washington suburb of Silver Spring, Maryland.

Although Mimi's mother never admitted it, she had another career in addition to housewife. Alma had attended two of the best art schools in the country: the Pratt Institute in Brooklyn and the Cooper Union for the Advancement of Science and Art in New York City. People paid her to paint their portraits.

"My mother worked in an art studio at home," Mimi marvels. "Yet she always denied that she was a professional—even though she sold her paintings."

Mimi's mother *(right)* captures a landscape on paper. A trained artist, Alma Koehl passed her artistic talents along to her daughter, Mimi, shown below at age 7 beneath her mother's self-portrait.

Besides portraits, Mimi's mom liked to paint landscapes. Sometimes she brought Mimi with her on hikes into the woods. "I'd set up my little folding chair and watercolor set," remembers Mimi, "and paint alongside her."

Her mother also took Mimi to art galleries in Washington, D.C. Mimi learned a lot from her mother, who could spend an entire day looking at the way other artists painted hands or the folds in fabric. Mimi especially liked Impressionist paintings in which artists made pictures from broken or flickering brushwork.

"When you stood up close to the paintings," Mimi explains, "they just looked like blotches of color. Then you'd walk away, and they'd turn into cathedrals or some scene."

Mimi's brother couldn't draw a straight line, but he noticed his sister's art skills right away. "Maybe it's because she's dyslexic," Bob speculates. "But Mimi had a gift for seeing things in a special way. If you had asked me what my sister would be when she grew up, I might have guessed 'artist.' I never would have said 'scientist.'"

~ A Growth Spurt for Squirt

Mimi was in the third grade when Bob went off to college. He graduated from high school two years early, at age 16. When he left, Mimi felt like an only child.

She enjoyed just a few more years of being a "squirt." In sixth grade Mimi went through a growth spurt that still makes her blush. "It happened too early for me," she says. "I was shy but I shot up fast and stood out—a tall, skinny girl still in elementary school. Most boys barely came up to my shoulders."

At summer camp in 1961, the 12-year-old Mimi *(far right)* had already begun to tower over her friends. She was headed for an ultimate height of 5′ 10″.

By junior high school Mimi felt like an ugly duckling. She loved taking modern dance lessons but her mother made her dance in a different way. Mimi's mother forced her to attend school dances with other teenagers. The events were pure agony: "My long, skinny body had trouble holding up the stockings and strapless dresses that were all the rage at the time," she recalls. "The only thing I held up was the wall. I just leaned against it, hoping someone would ask me to dance. No one did."

Even as Mimi grew out of her awkward early teens, she couldn't shake her negative self-image. "I was always surprised when someone asked me out," she recalls. "Now when I look back at myself, I realize the image other kids had of me was probably not as bad as the image I had of myself."

Mimi may have blended into the woodwork at school dances, but she stood out on the softball field. At a time when many girls shied away from sports, Mimi won a spot as catcher on the junior high girls' softball team. She explains, "I got picked because I was the only girl willing to mess up her hair with a catcher's mask."

~ *Undercover Rebel*

Mimi felt insecure for another reason. Her brother left behind a wall full of academic awards, which her parents bragged about. "He was a tough act to follow," admits Mimi, who went to the same public schools as Bob.

Mimi enjoys a close moment with her father, George, in 1960. A college professor of physics, George Koehl unlocked the door to mathematics for his daughter.

Because Mimi read slowly, she had to work extra hard to do well in subjects like history and English. Mimi's mother worried about the amount of time Mimi spent on her homework. If Mimi looked smarter than the boys, her mother warned, she might not get dates. Her mom fixed the "problem" by limiting study time, but Mimi read so slowly she couldn't complete her assignments.

Mimi rebelled. She crept out of bed early in the morning to do homework while her parents slept. One day her mother caught her in the act of reading. A neighbor who got up at dawn saw Mimi's bedroom light burning and told her mother.

"I got in all kinds of trouble," laughs Mimi. "My mother simply did not want a bookworm for a daughter."

Mimi's rebellion continued. She still got up early to study, but now she crawled under the covers to read with a flashlight. Mimi's efforts paid off: She got placed in the most challenging academic track at her high school.

Mimi's father taught her math. He spent long hours at the kitchen table, showing her how to organize equations so she wouldn't make mistakes. Mimi's love of math and science started at this table.

"It was really fun," she smiles. "He began with simple arithmetic in elementary school and worked with me all the way through high school. He showed me how math was a great tool to help me answer questions."

~ New Kind of Cool

Mimi might be shy but she didn't shy away from "uncool" activities. "I got bored with girls who wanted to talk about nothing more than boys or cars or makeup," explains Mimi. "I liked brainy kids like me."

Her mother pressured Mimi to do things that the popular girls did, such as becoming a cheerleader or a majorette. Mimi picked majorette.

"I practiced twirling and tossing a baton for hours," she recalls. "When tryouts came, I had to toss a twirling baton from one hand to the other. But I forgot that my head was in between, so I hit myself in the jaw and dislocated it." With a laugh, Mimi adds, "I didn't get picked. Worse yet, my jaw clicked for eight years after the accident."

According to Bob, boys didn't consider Mimi uncool. "Whenever I came home from college," he says, "there'd be guys lined up at the door."

Although Mimi "went steady" with three boys in high school, her mother never stopped worrying about her marriage prospects. No matter who she dated, her mother would say, "Your boyfriend is wonderful. Be sure you hang on to him."

But Mimi kept her sights fixed on college—and did a good job of getting there. In 1966 she graduated 18th in a class of more than a thousand students.

"That was the same class rank my brother had held," Mimi points out. "Yet my parents made me feel like he was smart and capable, but I was not. They encouraged him to go to Swarthmore College in Pennsylvania, but told me I couldn't succeed there—that it would be too hard for me. Isn't that weird?"

Not from Bob's point of view. "My sister struggled so hard with her studies, my parents wanted to protect her," he offers. "But she was already a star. My parents and my sister just didn't know it yet."

Mimi flashes her sunny smile at her junior prom in 1965 *(left)* and just before a date during her senior year in high school *(above)*. According to friends, Mimi's strong sense of humor has seen her through some tough times.

"The *forms* I found in *nature* inspired the drawings

and watercolors I created in the art studio."

3

The Art of Science

Mimi didn't know what lay ahead, but she couldn't wait to get to college. Her face brightens when she talks about her years at Gettysburg College in Pennsylvania. "I had roommates. I had friends. I went to parties. It was one of the happiest times of my life!"

She started off by enrolling as an art major. Mimi still questions her art talents, but her friends don't. "When I first saw her draw," says her friend Ellen Daniell, "I was totally surprised at how beautiful her sketches looked."

Mimi bowed to her parents' wishes and also signed up for elementary education classes. She did well enough in them to win the prestigious Columbia University Teachers College Book Prize in 1970.

Mimi's trademark long hair is evident in her senior yearbook picture *(above)* at Gettysburg College. After starting as an art major—that's one of her sketches on the opposite page—Mimi graduated with a major in biology. Throughout her life, her artistic eye has influenced the way she views science.

~ *Flower Child in Bloom*

While at Gettysburg, Mimi blossomed as a flower child. She adopted the "Earth mother" look: miniskirts or bell-bottom jeans, love beads, and handmade leather sandals. She wore her hair long, with big hoop earrings dangling from her ears.

Students gather for a discussion near Gettysburg College's Pennsylvania Hall *(right),* already 130 years old by the time Mimi arrived on campus.

At college, Mimi loved late-night discussions in the dorm about things like politics and science. She also loved going to parties, where she danced rather than holding up the wall.

Mimi was a party girl with purpose; she never let her grades drop. Gettysburg, a liberal arts school, required its students to take a wide variety of subjects so they would get a well-rounded education. Mimi liked them all, with the exception of French class. "I have a really hard time with languages," admits Mimi, "probably because of my dyslexia."

Mimi found the intellectual challenges exciting. "Like me, almost everybody else studied a lot," Mimi notes.

~ *Making a Difference*

Mimi's college years had a deadly serious side, too. She came of age during the Vietnam War (1963–1975), a time when many young people went off to fight in Southeast Asia. The war split the nation in half. While some people wholeheartedly supported it, others—often within the same family—bitterly opposed it. Mimi's family, by contrast, was of one mind on the issue. Her parents, especially her mother, opposed the war. Before heading off to college, Mimi had watched her mother regularly join protest marches in Washington, D.C. Her father, a college dean in that city, kept a low profile in public but spoke out against the war at home.

Mimi blossomed during her freshman year of college (1966–67). "I was free to reinvent myself," she said.

When Mimi got to college she joined the antiwar movement. Her parents did not object. "You couldn't escape the war," recalls

Mimi. "It hung over our futures like a big dark cloud. Back then guys were drafted into the military and had to go to war."

For a while Mimi's brother Bob temporarily escaped the cloud's shadow. He got a deferment, or official delay in military service, because he was a student at Yale Medical School. But what would happen to him after he graduated? Nobody knew for certain. So it was a huge relief for the entire family when Bob, now a doctor, got assigned to the Walter Reed Army Medical Center in Bethesda, Maryland.

Like so many young people at the time, Mimi felt a mix of fear and anger at what was happening in the world. "A lot of young guys I knew were drafted right after graduation and sent off to war," she says with a somber face. "Some got killed over there. Others came back with horrible wounds. Still others suffered from post-traumatic stress." (That's a condition in which they relived the war in their minds or dreams.)

As the antiwar movement grew, Mimi was motivated by the belief that her generation could make a difference. "If we got enough people involved," she explains, "we believed we might be able to change the world. We felt angry and afraid, but we did not feel powerless. We would make the country's leaders hear us—or so we hoped."

Gettysburg College students protest U.S. involvement in the Vietnam War in this 1969 photo from Mimi's yearbook, the Spectrum. Mimi took part in this demonstration and many others like it in Washington, D.C.

Recalling those turbulent times today, Mimi says, "We also worried about how people were destroying the environment. The first Earth Day happened on April 22, 1970, right before I graduated from college."

Mimi pauses. "I hope today's kids still think about these things—about war and the environment. I hope they still care. That's why it's so important to look back at the '60s and '70s; those years prove that young people can still do something about the really big issues."

Fascinated by forms in motion, Mimi captured these running figures in her freshman studio-art class.

~ A Major Decision

Amid this political turmoil, Mimi experienced some personal upheaval when she switched majors. Mimi liked art, but in college she discovered that she liked biology even more.

"I like art because it involves shape, size, and color," Mimi explains. "But biology looks at those things, too—it just looks at them in a different way. The forms I found in nature inspired the drawings and watercolors I created in the art studio. In biology I could learn how living things—the natural forms I liked to draw—actually worked. I wanted to know what difference it made to be big or small, to be round or flat, to be squishy or rigid. These questions interested me far more than painting."

Mimi's parents were unhappy with her decision. Even her brother seemed taken aback. "Mimi always did well in math and

Mimi's painting of a smoke-belching factory contrasts with her graceful rendering of a cow skull. Biological shapes like the skull drew her ever deeper into science. Today Mimi draws many of her own scientific illustrations.

science," Bob explains, "but she had never spoken about a career in either subject."

"I don't think anybody took me very seriously, not even my professors," says Mimi. "I got good grades but I had missed a lot of science courses while pursuing that art degree. I had no calculus, no organic chemistry, no genetics, no physics—none of the special courses needed for graduate school."

Despite the odds Mimi stuck by her decision. Science came so naturally to her that she started tutoring friends who struggled with the subject. "I realized that I liked taking apart complicated problems and explaining them in words anybody can understand," she smiles. "But it never entered my mind that I might enjoy being a college professor. That was a big leap of the imagination—not just for me, but for any woman who majored in science during those years."

~ *Falling in Love*

Another notion that hadn't seriously entered Mimi's mind at the time was marriage. But in June of 1968—the end of her sophomore year in college—one of Mimi's friends asked if she wanted to attend "June Week" at the United States Naval Academy in Annapolis, Maryland.

June Week was filled with parties, formal dances, and all sorts of other events. "My friend was going with a cadet at the academy," Mimi remembers, "and he had some friends who needed dates for June Week. So I said, 'Sure!'"

Mimi and her first love, Michael, relax on a date in 1968.

At one of the parties Mimi met a naval cadet named Michael. "I immediately recognized him as someone special," she recalls. "Michael wore the white uniform of an officer-in-training, yet he questioned the Vietnam War. He was attending the Naval Academy to get a top-notch science education."

"We fell madly in love," Mimi continues with a smile. "But Michael didn't know where he'd serve, if he'd fight, if he'd live or die."

After struggling with his conscience over whether he could command in a war he didn't support, Michael dropped out of the Naval Academy. To fulfill his obligations to the Navy, he had to serve as an enlisted sailor aboard an aircraft carrier. Eventually, he received a discharge.

She received her diploma wearing a dress—and sporting a peace band wrapped around her arm.

When Michael told Mimi that he planned to attend graduate school, she applied to many of the same schools. "He got accepted," she says. "But without enough science, I got turned down everywhere. Only Duke University in North Carolina showed any kind of interest: It put me on its waiting list."

The love story didn't end, though. "A lot of our friends were getting married as soon as they graduated from college," continues Mimi. "Michael and I planned a June wedding, too."

Rather than renting a cap and gown for her Gettysburg graduation in May 1970, Mimi joined many of her other classmates in donating that money to the election campaigns of peace candidates to Congress. She received her diploma wearing a dress—and sporting a peace band wrapped around her arm.

Then Mimi dropped a bombshell. Three weeks before the wedding she canceled it. "I wasn't ready to stop learning," she states with conviction. "I couldn't just follow Michael around. So my first great love and I went our separate ways."

In her senior year of college, Mimi's classmates voted her homecoming queen. The formal portrait above was taken to celebrate that honor.

~ *Charting a Course*

While Michael headed off to graduate school, Mimi went to the Woods Hole Oceanographic Institution on the coast of Massachusetts. It's the largest independent marine research center in the world. Woods Hole offered summer fellowships to students interested in oceanography careers. Mimi didn't win a fellowship, but she did land a job as a lab technician.

Mimi's task was to help a scientist study the effects of industrial wastes being dumped by a paint company into the waters off New Jersey. She did the chemical tests needed to determine which dangerous materials marine organisms might be picking up from the fouled water. "I learned all these new techniques," Mimi says, "but I got frustrated working on somebody else's research questions instead of my own."

Woods Hole Oceanographic Institution in Massachusetts, where Mimi landed after graduation, boasts 70 years of seagoing experience. Since 1930, it has lured students like Mimi to explore the movements and life-forms of the ocean.

Toward the end of the summer Mimi remembered that Duke University never sent her a rejection letter. *Maybe I'm still on the waiting list,* she thought. Mimi summoned her courage and phoned the director of graduate studies. The conversation went something like this:

"Hello. This is Mimi Koehl. I applied to your graduate program in zoology, and you put me on the waiting list. But I never heard back from you."

After shuffling and rustling some papers for a minute, the director replied, "You're still on our waiting list, but I see that somebody just dropped out. Why don't you take that spot?"

Mimi couldn't believe her ears. Duke had one of the best zoology programs in the country. Its professors did groundbreaking work in the study of animals, including the sea creatures that Mimi loved.

~ *What If?*

As Mimi re-creates that conversation today, her eyes get wider by the minute. Shaking her head, she says, "Suppose I'd never gotten the nerve to make that call. Can you imagine where I might have ended up in life?"

Mimi's friend Ellen Daniell doesn't even like to think about that possibility. "Those of us who know Mimi Koehl believe it would have been a crime for her not to go into science. She gets so much joy from it. She wants to share it with everybody."

In the fall of 1970 Mimi charted a new course for herself. This one led her into the ocean—and deep into a new science called biomechanics.

"So that's how

I started graduate school—

with a bang!"

4

THE MENTOR

Mimi made a real crash landing when she arrived at Duke University to study marine zoology in the fall of 1970. After several nights in a sleeping bag on other people's floors, she moved into a house jammed with near-penniless graduate students like herself. There was no bedroom for Mimi, so the students had to create one. They erected a flimsy plywood panel that divided one small bedroom into two tiny ones. A curtain served as her door.

"My half of that space was so small," Mimi recalls, "that I couldn't fit both a mattress and a chest of drawers in there. So we went out and found a tall, narrow lab bench in a trash dumpster on campus. It had drawers for my clothes in it, and I thought it would make a good bed platform. We wedged it into my room, then topped it with a board and a mattress."

When Mimi's alarm went off the next morning, she leaned over to turn it off and upset the delicately balanced bed. With a crash and a smash, the mattress flipped her into the makeshift wall, creating a sandwich with Mimi in the middle.

The graduate student on the other side of the partition awoke with a yowl. He knew the room divider might topple onto his head at any second.

Top science students pass through the doors of the Biological Sciences Building *(opposite)* at Duke University in Durham, North Carolina. Here in the 1970s Mimi met some of the pioneers in biomechanics. They inspired her to explore life-forms along rocky, wave-swept sea coasts *(above)*.

"The entire house ran to my little cubicle to see what had happened," laughs Mimi. "All they saw were two sets of toes wriggling between the mattress and the plywood wall. So that's how I started graduate school—with a bang!"

~ *One of the Guys*

Mimi lived with male housemates because that's who made up most of the science graduate program. Everybody took turns cooking, but Mimi resisted it at first, because it was a "girl thing"—and because she had never learned how to do it. That all changed the day she served her housemates a meal of burned hot dogs. "The guys taught me it was okay for a woman scientist to know how to cook," Mimi jokes.

It didn't take Mimi long to find out where all the graduate zoology students, including a handful of women, hung out. "They had this funky seminar room with a big old couch and a coffeemaker," says Mimi. "We'd go there and sit around talking about science and about our research."

Mimi had found a home. Each morning, the zoology grads would help one another with technical problems. They scribbled equations on a chalkboard in the room. They argued, waved their hands to make a point, spilled coffee on their notebooks, and told dumb science jokes—often all at once. "I probably learned as much science in that coffee room," Mimi reports, "as I did in any of my graduate classes."

Mimi lived with male housemates because that's who made up most of the science graduate program.

~ *Make-Up Lessons*

It took Mimi time to loosen up with her classmates. The confidence she gained at Gettysburg had been shaken by Duke's late acceptance. As Mimi puts it, "I arrived thinking, I must be the dumbest kid in the class—the bottom of the barrel, the last one chosen for the very last spot. It felt so scary. At the same time, it was exciting."

The director of graduate studies told Mimi that she had to take all the science courses she had missed at Gettysburg during her first year at Duke. "This would have made any other graduate student I've ever known drop out of school," says Steve Wainwright, who later became Mimi's mentor at Duke. "But Mimi just buckled down and did it."

In addition to taking classes, Mimi earned her living by working as a teaching assistant in an introductory biology course.

Mimi wears what she jokingly calls her "hippy-dippy" clothes to a picnic for the Duke University zoology department in 1972. To pinch pennies, Mimi made many of her own clothes, including these once fashionable pants.

One day she walked into a class that changed her life. At the front of the room, Steve Vogel, a pioneer in biomechanics, talked about the physical and mechanical consequences of being very big or very small in nature.

Mimi, who had pondered the very same question at Gettysburg, couldn't wait for the lecture to end. She went up to Vogel afterward and said, "That's some of the neatest stuff I've ever heard. Where can I learn more about it?"

"My colleague Steve Wainwright is the expert on this subject," said Vogel. "But he's away on sabbatical [academic leave] doing research. So you'll have to wait for him to get back."

Mimi resolved to approach Wainwright when he returned to campus in the fall of 1971. In the meantime she flung herself back into basic science.

~ No Women Allowed

Mimi still thought about becoming an oceanographer. "Oceanographers combine physics with biology," she explains. "They use physics to look at how an ocean moves and biology to study what happens to organisms that live in the moving water. It brought together fluids and natural forms. As an artist, I saw movements and shapes as beautiful things."

To explore oceanography some more, Mimi applied a second time for a summer student fellowship at Woods Hole. This time she won it. "I was no longer a lab tech," she points out. "Now I actually got to do my own research project in somebody else's lab."

The other fellowship winners were men. They lived in a dormitory at Woods Hole, while Mimi rented a room in town. "I felt left out," she remembers. "They got to spend nights talking about their science or having fun together."

Yet Mimi faced a bigger problem. At the time, Woods Hole had a rule barring women from its oceangoing vessels. So while she could design an oceanography experiment, collect her gear, and load it on a ship, she couldn't go to sea. Mimi had to persuade the men to conduct her oceanography experiment for her.

"If you're lucky, like me, your major professor becomes a wonderful mentor."

"I didn't want to stand on a dock while someone else did all the fun stuff," Mimi exclaims. "I wanted to do my own fieldwork." The big question was: If not oceanography, then what?

~ Soul Mates

For an answer, Mimi sought out Steve Wainwright, who had returned to Duke by the fall of 1971. She expected to find him in a cubbyhole crammed with books, like the other professors. Instead, she says, "I saw pictures, a big lime-green couch, and wall hangings. There was wonderful artwork everywhere."

Mimi recognized Steve as a fellow artist. She was right. In addition to collecting art, he created it as a sculptor. And because Steve had studied Mimi's records, he knew all about her own art background. "We're both visual people," explains Mimi. "So we see the world in terms of shapes and sizes and movements. We understood that about each other right away."

Mimi didn't notice it at first, but Steve kept a box of tissues by his lime-green couch. Graduate students, both women and men, trusted him so much that they often came to him and poured out their problems. When Steve saw tears coming, he would offer tissues along with some advice.

Mimi, who soon became Steve's pupil, was no exception. As she unburdened herself to Steve, he discovered that this highly gifted scientist had a low opinion of herself. As Steve puts it, "Her self-esteem was lower than whale droppings on the bottom of the ocean floor." Mimi laughs when she hears this. "I'd say that's about right."

Steve Wainwright stands behind his wife, scientist Ruth Palmer, in this portrait from the mid-1970s. Ruth, often called upon to feed hungry graduate students, was remembered by Mimi as "funny, lively, a good friend, and the world's best cook."

Steve agreed to be Mimi's Ph.D. advisor, but with one condition: He required all his graduate students to take a tutorial with him. "It's the hardest work you can ever do," Steve observes. "You have to write five papers in a semester, and each one gets examined down to the letter. Mimi knew the rule, and she took it on."

"I really wanted Steve to be my major professor," says Mimi. "That's the person who teaches you how to formulate research questions; how to figure out if those questions are worth answering; how to design experiments that will answer the questions; and how to test your answers. If you're lucky, like me, your major professor becomes a wonderful mentor."

~ Of Crystal Balls and Squishy Things

Steve frustrated his graduate students with something he calls the "crystal ball technique." Mimi explains how it works: "I'd run into his office all excited about some research question, and Steve would say, 'If I had a crystal ball in my hand right now and it showed me the answer to your question, so what? What would you know?'

"He never stopped bringing out that crystal ball. He wanted us to look at the big picture—to develop scientific principles that no one had ever thought of before."

Steve specializes in invertebrate marine zoology, the study of sea animals without backbones. So Mimi's tutorial got her really interested in what she calls "squishy things." Mimi recalls her surprise upon realizing that animals without bones actually do have skeletons, called hydrostatic skeletons. "I had never really thought about how organisms operate without a bone in their body. Think about it: A worm can make itself long and skinny, or it can contract itself into a fat wad. It can even push its way through the ground!"

"I was burning to find out the physics behind these behaviors. Why? Because the ability of a squishy thing to exert force on its environment is one of the most intriguing feats I'd ever run into. I mean, how can a soft lump dig a burrow in hard ground? When I realized that a hydrostatic skeleton is basically a water balloon with muscles, I decided to write one of my tutorial papers about how it works."

Because of Steve's cutting-edge research in biomechanics, Mimi started to think like an engineer. She asked structural questions: How are squishy skeletons built? Why are some squishy skeletons tough and strong while others are soft and stretchy? How does any squishy skeleton help a given creature survive in its environment?

Mimi gave a lot of thought to the organism that might best help answer her questions about hydrostatic skeletons. Her search ended when Bob Paine, a marine ecologist at the University of Washington, visited Duke. He gave a talk about his study of the ecology of the animals and algae that inhabit the wave-battered coast of the Pacific Northwest.

Mimi came alive when Bob showed slides of big, beautiful, bright green sea anemones [uh-NE-muh-nees]. "I knew about the delicate, fluffy anemones that live in calm places," she explains. "But I had no idea that these large, flowerlike creatures could live among the crashing waves. They're nothing but water balloons—yet they're beautiful and they're tough. These were the organisms for me!"

When the lecture ended, Mimi raced into Steve's office. "She looked at me with eyes as big as saucers," Steve recalls, "and said: 'I know what I'm going to do for my Ph.D. I'm going into the ocean to look at sea anemones. I'm going to use them to figure out how hydrostatic skeletons work!' "

With not a bone in their bodies, giant green sea anemones look more like flowers than animals.

Steve didn't need a crystal ball to know that Mimi had found her place in science. Nobody had ever put on a diving suit and gone eyeball-to-eyeball with a sea anemone to find out how this living machine worked. Until now.

They act like baseball mitts,
catching mussels
and sea urchins

knocked down
on them by waves.

5

FRIENDS AND ANEMONES

Mimi's Ph.D. research took her into a rough neighborhood, one where humans can take a real pounding if they're not careful. Imagine standing up to waves taller than you. Imagine, too, holding your ground against waves moving at top speeds of up to 45 feet per second. And how about staying put with all those waves sloshing back and forth? This is not an easy place for people to get a firm footing.

A giant green anemone *(opposite)* doesn't hunt. It catches food—mussels and sea urchins—knocked into it by waves. Mimi studied these creatures on the island of Tatoosh, Washington *(above)*.

Yet a surprising number of plants and animals make their homes along wave-swept, rocky seacoasts like this. Many of them, including squishy sea anemones, cling to the ocean bottom close to shore without getting ripped away. Mimi wondered, *How do they manage to do that?*

To answer to this question, Mimi put on her engineering hat. She looked at giant green sea anemones as an engineer might study an airplane or a bridge. She wanted to see how the design of these living machines let them withstand—and even use—the water roaring past them.

Mimi knew the ideal place to study the animals: Tatoosh, an island off the northwest tip of Washington State's Olympic Peninsula. That's where University of Washington ecologist Bob Paine had been doing research since the 1960s. Bob calls the island's rich tidal system a "paradise of biology."

The Makah, a Native American tribe, manage the 20-acre island of Tatoosh *(above)* but they don't live on it. Their reservation is the town of Neah Bay on the Olympic Peninsula. The inset shows young girls in tribal dress during Makah Days, one of the tribe's annual celebrations.

Mimi wanted to visit Bob's outdoor "laboratory," which he was using with the permission of the Makah, the Native Americans who still manage the island today. So in the spring of 1972 she wrote him a letter.

"I asked if I could tag along on one of Bob's summer field trips with his graduate students," Mimi recalls. "While they worked on their ecology questions, I could work on my own questions about sea anemones."

"Sure," Bob wrote back. "Come along."

~ *A Ride in a Rocking Bucket*

Mimi describes Bob Paine as "bigger than life." At 6' 7" tall, he towers over most people. But he also cuts a tall figure in ecology and marine zoology. "If I had known how famous he was," admits Mimi, "I might never have gotten up the nerve to write to him."

Mimi didn't realize that Steve Wainwright had given her a glowing recommendation. However, Steve believed Mimi faced some danger. As he puts it: "I knew she'd be heading into the surf-beaten, intertidal zone—the area between the upper and lower tide marks. The currents there can be strong, and really big waves,

Bob Paine holds a starfish he pulled from fast-flowing water on Tatoosh, the outdoor "laboratory" he opened to Mimi. Bob established the principle of the "keystone species"—an organism critical to the balance of an ecosystem. The starfish is the keystone species on Tatoosh.

created by faraway storms, can come out of nowhere. This is high-risk work."

With a laugh, Bob Paine adds, "Back in those days—and even now—just getting to Tatoosh was an adventure."

The scientists boarded a Coast Guard cutter at a station in Neah Bay. After bouncing through choppy seas, they arrived at Tatoosh, with cliffs rising 40 feet above the ocean.

"Back then people ran the island's lighthouse, which is automated now. The lighthouse workers used a crane at the top of a cliff to lower a big bucket down to the boat. We'd climb into the bucket, and they'd haul us up," explains Mimi.

"On rough water, it got really interesting," grins Bob. "The rocking bucket felt like an amusement park ride, only scarier—and with real crashing waves."

~ "Something Is Wrong!"

Once Mimi stood safely atop the cliffs, she could see and hear fierce ocean waves bashing the rocks below. She had to get down to that battered shore, however, to measure how fast the water was moving across the giant green sea anemones that lived there. Safety required two conditions: low tide and calm weather.

Mimi climbed down a path skirting the cliffs. Then she stepped, slipped, and slid along rocks covered with gooey algae and sharp barnacles. Lugging equipment—scaffolding, bolts, electronic instruments, and cables—made the trip even harder.

As Mimi tells the story, "I needed to hook up electronic devices to measure water speeds and the physical forces felt by the animals. When the tide was out and it was safe to visit the sea anemones, I spent hours bolting equipment to the rocks, waterproofing gear, and running and tacking down cables that stretched to a safe spot above the high-tide mark."

Low tide exposes the coastline below the cliffs of Tatoosh *(above),* allowing Mimi *(right)* to lug a backpack full of gear across the algae-covered rocks. She used the knife strapped to her left leg to scrape away algae so she could bolt equipment to the rocks.

Mimi didn't have a laptop computer—its development was still years in the future—so she relied on devices that sent data in high-pitched and low-pitched electronic sounds. Mimi recorded the weird squawking on a boom box, then analyzed the squawks back in the lab.

Even with such hard work, marine experiments can fail. Waves knock cables loose. Ocean salts short out batteries. Waterproofed equipment gets ripped apart. So when Mimi studied her early readings, she thought something had gotten messed up. "The waves were really big," declares Mimi. "But when I put my equipment down with the sea anemones, it sent back these wimpy signals. I thought, 'Oh, no! Something is wrong!'"

The equipment was working fine—it's just that Mimi had been looking at the sea anemones' neighborhood from her own perspective. Mimi saw turbulent waves. But that's not what the sea anemones saw. In wave-exposed habitats, they flatten themselves into one-inch-high pancakes. This shape helps them hide in the boundary layer—a layer of water close to the bottom that moves more slowly than the water rushing overhead.

The anemones in wave-tossed habitats use another survival

trick, too: They hunker down behind their neighbors, sheltering themselves even more. "It wouldn't be a good idea to be taller than your buddy," laughs Mimi, "because then you'd get walloped by the faster water flow."

~ *Standing Tall*

Like the best of scientists, Mimi Koehl doesn't mind proving herself wrong. She throws out a theory if the data disproves it and comes up with a new theory to test.

And so it was with the giant green anemones. The animals, Mimi discovered, didn't stand up to the waves; they ducked them. So she wondered, *How does the structure of the giant green anemones on wave-swept shores differ from the structure of their relatives in much calmer sites?*

To human eyes the world looks safer in protected bays with gentle currents than in wave-blasted surge channels. But Mimi got another surprise: She learned that white anemones in protected sites stand tall, holding their crowns of tentacles 12 or more inches above the bay's floor. So these calm-water anemones can actually get hit with faster water speeds than the green anemones, which tuck themselves into the slower moving boundary layer.

Surf roars into the surge channel, or rocky inlet, where Mimi set up equipment to measure the waves hitting sea anemones in 1974.

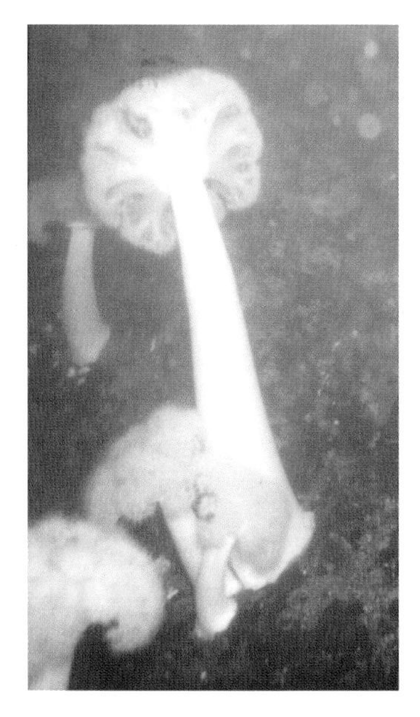

Unlike its flatter, rough-water cousin, this calm-water sea anemone, known by the scientific name *Metridium senile,* is made of a tissue that bends with tidal currents.

At protected sites the ocean current flows in one direction when the tide goes out and in the opposite direction when it comes back in. As the tides flow past the tall anemones, the gentle currents gradually bend over the stalks until their crowns of tentacles end up perpendicular to the water flow. This allows the tentacles to catch plankton as the water moves through them.

The flowing water, however, exerts a much greater force on the anemones in this broadside position. To compensate they have a built-in safety mechanism: When the ocean current gets too fast, the anemones' tentacle crowns simply collapse, like umbrellas flipped inside out by a gust of wind. The water's "drag" is much lower on these collapsed, streamlined bundles of tentacles.

~ A Tissue Issue

All this bending and stretching led Mimi to think about body tissues. As she explains, "Sea anemones are basically water balloons. An outer bag—made of special material called mesogloea [me-zuh-GLEE-uh]—surrounds a water-filled gut. I wanted to know how the mesogloea of the white sea anemones, which stretch and bend in the water flow, differs from the mesogloea of the giant green anemones, which don't get stretched by the waves."

The question preyed on Mimi's mind. She carefully cut slivers of tissue from anesthetized anemones. (The animals healed quickly and lived for years.) She put the slivers in a cooler and carried them across campus to the engineering department. There she could pull on the mesogloea in a hulking machine called an Instron. The machine measures how hard a material must be pulled to get it to stretch or break.

"The engineers didn't expect somebody to test tiny, thin chunks of sea creatures in their lab," Mimi declares. "Plus they couldn't see anything in this big machine. I looked up and saw all these eyes staring at me."

Mimi's tests showed that anemone tissue acts like Silly Putty, the children's toy. "Both mesogloea and Silly Putty are hard to stretch if you yank on them very quickly, like a wave," Mimi says. "But the two materials can be stretched easily if you pull on them gently for a long time, as tidal currents do."

What a Drag!

Drag is what slows down parachute jumpers as they fall through the air. It also slows down divers as they plunge into a pool and flying animals when they spread their wings to land. Drag is the resistance of a fluid—air or water—to an object in motion. The bigger the exposed surface, like spread wings, the greater the resistance.

Suppose you're rooted in place and fluid moves past you, much like the case of a tall, calm-water sea anemone. Would you feel drag? Try this to learn the answer: On a blustery day, hold a piece of stiff cardboard perpendicular to the wind. Drag is the force of the air pushing your cardboard in the direction of the wind. In the ocean, drag is the force of water current pushing on the sea anemone as it stands in place.

Now hold the cardboard so that it's parallel to the direction of the wind. You'll find that the drag is much less. Tightly fold up the cardboard and hold it up to the wind. How strong is the drag now?

Look at the diagrams showing the postures of a tall sea anemone. Its crown of feeding tentacles is perpendicular to the flow direction in drawing A, so the drag on it is large. But when the current speeds up (drawing B), the flexible animal is blown into a more compact shape. What would this shape change do to the drag?

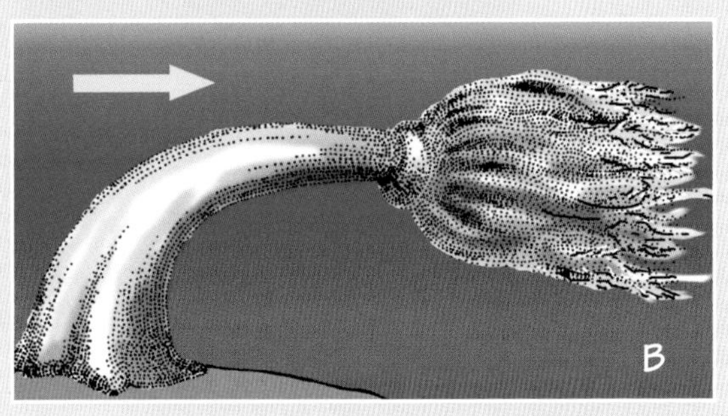

Mimi found that the mesogloea from the white anemones stretched much farther than the mesogloea from the green ones. Based on still other complicated lab experiments, she learned that the two types of mesogloea have different molecular structures. Mimi had used the sea anemones to learn what makes some types of tissues harder to stretch than others.

"I got really excited by my findings," declares Mimi. "I like to know how organisms work. I had discovered how the mechanical behavior of sea anemone tissue helped the animals catch their meals from the ocean.

"The calm-water anemones can be bent by currents into a position that helps them filter plankton as it drifts past their tentacle nets. They don't snap back quickly. The wave-tossed anemones, with more rigid bodies, don't fish around in the upper currents. They act like baseball mitts, catching mussels and sea urchins knocked down on them by waves."

~ Model Behavior

"What if?" is a question that Mimi asks a lot. She used it to learn more about other body features of the two types of sea anemones. What if the giant green anemones weren't covered with wart-like bumps? What if the smooth white anemones did have these bumps? What if the two animals weren't coated with slime? What if each species was a different shape or size? What if neither species had tentacles?

Mimi wanted to know which types of body features had an important affect on drag. She couldn't use real sea anemones to test every single variable that interested her, so Mimi built models of the animals. This skill has become one of her trademarks as a scientist.

Mimi began by taking a pottery class near Duke, where she was working on her Ph.D. Here she made clay models in all different sizes and shapes. "It looked odd," Mimi laughs, "to see my anemones lined up for firing in the kiln alongside the mugs and bowls made by other students."

The sea anemone models above were some of the first "Mr. Potato Heads" that Mimi made. First she fashioned clay bodies, then she added or removed warts, tentacles, and other features. Which do you think are models of calm-water anemones? Of giant greens?

The ceramic models worked like Mr. Potato Head, a popular children's toy. Mimi could stick on or remove all kinds of variables, including warts, tentacles, or slime.

To test a particular variable—warts, for example—Mimi needed to place models with and without warts into flowing water and measure the drag. So she set up her own "ocean" environment.

Mimi asked Steve Vogel, one of her professors at Duke, if she could use his flow tank—a very long aquarium through which water flows. Controls on the tank let her set the water speed to mimic the flow experienced by different species of sea anemones. She then tested one variable at a time. Warts and slime didn't make much of a difference to how the animals interacted with the flow. But—as Mimi had discovered in the field—the creatures' size, shape, and orientation to the current did matter.

~ *Going with the Flow*

Is that the whole story? Hardly. In the early 1970s, while Mimi was still working toward her Ph.D., some new characters floated to the surface.

"I was doing my fieldwork on sea anemones in the coastal waters of the Pacific Northwest," Mimi exclaims, "when I noticed all these gorgeous seaweeds. They didn't hide like the giant green anemones. The ocean knocked them around, but they didn't break. These big leafy structures just fascinated me. I had to know how they worked."

There was just one problem with this detour on the way to earning a Ph.D.: Mimi's work almost swept her away.

Then, *with a whoosh,*

the **killer wave**

washed Mimi **away**.

6

Introducing Dr. Koehl

Wham! Mimi never saw the huge wave coming. It was the summer of 1976, and she was working her way around a channel of water cutting through the rocky shore at Viña del Mar, Chile. She crouched high and dry on a cliff, poking a long pole into the water. A funny-looking device strapped to the end of the pole measured the speed of the water rushing past the seaweeds that clung to the sides of the channel. Suddenly, a rogue wave—a lone wave formed when faraway storms push smaller waves together—slammed into the cliff. It scooped up Mimi and sucked her beneath its salty waters.

Pacific waves pound the coast of Chile *(opposite)*, where Mimi studied seaweeds in 1976—and nearly paid for her curiosity with her life. Above, a tower of wave spray dwarfs marine scientist Bob Paine and graduate student Rich Palmer.

Then, with a whoosh, the killer wave washed Mimi away. Fortunately it didn't carry her out to sea or hurl her into the rocks, as other rogue waves have been known to do. Instead it washed her back ashore.

"The next thing I knew," recalls Mimi, "I was soaking wet, sitting on another cliff without my equipment, and covered with cuts and scratches from the barnacle shells glued onto the rocks."

Off in the distance Mimi saw a bunch of people, including Bob Paine, waving their arms and sprinting toward her. Mimi had joined Bob on another trip with his graduate students to study marine life along the Pacific seacoast of South America.

Certain seaweeds survive—and thrive—on surf-churned stretches of coast like this one in Chile. In the summer of 1976, Mimi perched atop these cliffs to measure the speed of the water hitting the kelp.

From a marine lab nearby, Bob, his students, and some Chilean scientists had watched the near-disaster unfold.

"Parts of the Chilean coast are notorious for rogue waves," Bob explains today. "Some really fierce storms in Antarctica kick them up. We were sitting in the lab talking about science when we saw this wall of water heading toward Mimi. We watched helplessly. Luckily, Mimi didn't drown."

~ *Stretching Science*

Mimi had started studying seaweeds before heading to Chile. She got interested in them while researching the tall, slim, calm-water sea anemones found near the Friday Harbor marine labs.

Friday Harbor lies on San Juan Island, in the northernmost stretch of Washington State's Puget Sound. Here diverse seaweeds hug the shorelines. They vary in form from giant bull kelp with 100-foot-long stems (called stipes) to leafy, lettuce-sized seaweeds.

Mimi wanted to examine the relationship between water flow and the design of these plants. She also wanted to get Steve Wainwright out to Washington. As he remembers it, "She walked into my office in North Carolina and said, 'You, Steve Wainwright, have to go to Friday Harbor.' She kind of packed up my lab and three other graduate students and took us to study the mechanics of marine creatures. Of course, I was very excited. It was the birth of this kind of biomechanics."

Mimi and Steve talked a lot about the mechanical forces that work on solids. As Steve recalls, "Mimi and I got down to two basic kinds of forces: tensile forces, or ones that pull, and compressive forces, or ones that push. Tensile forces are easier to measure in the field. So we agreed to find an organism that functions in tension."

They settled on bull kelp. It attaches to the seafloor with a "holdfast" bigger than your hand. At the top of the stipe, meanwhile, a gas-filled float holds the long, leafy blades up in the light near the surface. From their study of physics, Mimi and Steve knew that the main forces working on the kelp would be the waves and currents pulling on its stipe. They came up with a prediction: The kelp would be made of a high-strength material able to resist pulling.

Gas-filled floats help kelp stand tall in deep water *(above).* Steve Wainwright *(left)* hunkers down on the deck of a research vessel during a foray from Friday Harbor. "I've always admired the polar-bear types who like this kind of weather," laughs Steve, who prefers Hawaiian shirts and the warm waters off the Carolina coast.

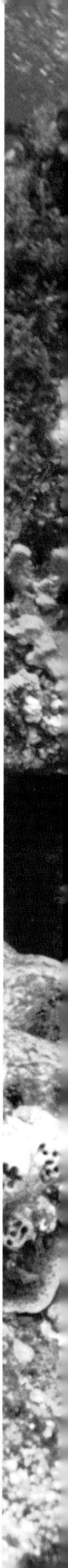

Mimi and Steve did their first tests. "We were really wrong," says Mimi. "It doesn't take much force to tear a piece of kelp. It's very weak."

"I don't know about Mimi," adds Steve, "but I didn't get much sleep trying to figure out the mechanical principle that keeps kelp from being ripped apart by the waves pulling on it."

Mimi and Steve can show you complicated numbers and graphs that explain the answer. But they also like to explain the kelp-wave dynamic in simple terms: The stipe stretches like a bungee cord, absorbing the force of a wave so that the kelp is not yanked hard enough to break. And just like a bungee jumper bounces back up after a fall, the rubbery kelp bounces back to its unstretched shape after each wave passes.

"We learned that if you're building a giant kelp, you want to use a stretchy material," says Mimi. "Also, in places where water moves back and forth in waves, it helps to be long. Before the plant ever gets fully stretched, the water usually starts flowing in the opposite direction. That means the holdfast never feels a really big force. It's kind of like walking a dog: You don't feel the force of the dog pulling on the leash until it stretches it out."

~ *Pickled Stipes*

Talk about luck. Just as Mimi was getting more interested in seaweeds, she found out in 1976 that Bob Paine had won a grant to study the exposed rocky coasts of South America. He wanted to compare the habitats there with those found on the Pacific Northwest island of Tatoosh. Knowing that Mimi was a hard worker, Bob offered her a chance to help. In exchange, she could spend some time on her own research.

Mimi jumped at the chance. "I wanted to look at the designs of two different seaweeds that lived in the same wave-swept habitats. One had stiff stipes, so the kelp stood up like little trees. The other had long floppy blades."

Luggage limits at the time barred Mimi from carrying much gear. And because the field of biomechanics was so new to science, the Chilean marine biologists didn't have any biomechanical equipment for her to borrow.

Mimi ended up crafting many of her own measuring devices. To see how much kelp stretched, she made baskets from wire screens that washed up on the beach. She weighed pebbles and rocks precisely, then painted the weight on each one. She attached the basket to a piece of kelp, hung it off a ledge, and dropped in the numbered weights. A partner measured how much it stretched with a tape measure. When Chilean kids walked by the pile of numbered rocks, they giggled, *"Los numeros!"*

> Mimi ended up crafting many of her own measuring devices.

Mimi also used a mechanical flow meter, not an electronic one, to measure water speeds. "I combed the beach for a long stick so

I could attach this device to it," Mimi explains. "Then I'd sit on rocks, lower it into the water, and wait while the moving water spun the propeller around and around. If the water moved faster, the propeller spun faster, too. Numbers on the propeller showed how fast the water was moving. That's what I was doing the day the rogue wave washed me away. It spit me back out, but it kept my flow meter."

Mimi's work in Chile showed once again the diversity of nature's machines. Some of the seaweeds she studied were floppy and stretchy, like the bull kelp; others weren't such pushovers. "One species," Mimi points out, "was really woody—kind of like a tree."

Marine laboratories dot the 3,000-mile-long Pacific Ocean coast of Chile. In the summer of 1976, Mimi and the Paine expedition started at the northern tip of the country and worked their way south.

As a biomechanist, Mimi wanted to examine the cellular structure of the stipes more closely. So she pickled pieces of different stipes—that is, she preserved them—and stuffed them into her backpack. Later on, back in the United States, she used an electron microscope to study how the fabric in various stipe walls was woven together: Did the material have densely packed cells? Or were its cells loosely packed? "That's how I figured out," says Mimi, "what made one kind of stipe stretchy and a second kind as stiff as wood."

~ Coasting Along

As Bob and his team made their way down the Chilean coast, they often slept on the floors of marine labs. They repaid their hosts' kindness with some good old-fashioned home cooking: spaghetti, garlic bread, and apple pie. "I can't say we introduced those foods to Chile," admits Bob, "but we put on a good feed wherever we went."

When the team got ready to leave Chile, Mimi cooked up a plan with Ken Sebens, one of Bob's graduate students. They decided to head across South America to the coast of Brazil, on the Atlantic Ocean, to look at the zoanthids—close relatives of sea anemones—that live there.

Zoanthids like to hang out in crowds, preferably in coral rubble. There, a single animal buds off copies of itself, forming a sheet of creatures connected to each other. "They're genetically the same," says Mimi.

Ken, an ecologist, wondered what the creatures eat. He also wanted to see how they interact with each other and with other species of life in the neighborhood. As a biomechanist, Mimi was interested in examining the design of these animals.

"So we decided to find out what these guys ate and then go our separate ways from there," Mimi recalls. "For me, that meant finding out how they catch the food from the water flowing past them."

Toting 40-pound packs, Mimi and fellow biologists Ken Sebens and Carol Slocum prepare to travel to a new research site on the Chilean coast.

Zoanthids, it turns out, eat plankton and other tiny particles carried by the water flowing by. But how do they catch them? Another graduate student, Dan Rubenstein, had already tickled Mimi's imagination with an idea. Sitting in the zoologists' coffee lounge at Duke, he had told Mimi one of his theories: Dan thought some animals filter particles of food from the water the same way a furnace filter screens particles of dirt from the air. So the two scientists wrote a paper using math to calculate how different filter designs would work.

The filter-feeding theories drifted around in Mimi's mind whenever she looked at animals that feed on plankton. She asked herself, *How does the structure of that filter affect what it can catch? How does it work?*

"The zoanthids in Brazil offered a chance to study how the size and shape of a group of animals affect their ability to filter plankton," Mimi explains. "Plus there were so many different kinds of zoanthids to study. It was great!"

Because Mimi had taken time off from her sea anemone project to join the trip to South America, she had to solve a big problem: Somehow she had to write her Ph.D. dissertation—a book about her research findings on sea anemones—in record time.

"Graduate students sweat and bleed over their dissertations," explains Mimi. "They take forever to write them. I had just two months to get it done before starting a postdoc position at Friday Harbor."

This colony of anemone-like zoanthids formed when one individual budded off copies of itself. After studying zoanthid colonies in Brazil, Mimi and Ken headed to Panama to repeat their experiments.

~ *Ticket to the Future*

So after spending the summer in South America, Mimi hustled back to Duke. She decided to title her dissertation *Mechanical Design in Sea Anemones.* Each chapter would answer a specific engineering question about the complex design of these simple creatures.

In that precomputer era, Mimi wrote out everything in longhand, then handed the pages to a typist. While the typist click-clacked away, Mimi went into a darkroom to develop the photographs of sea anemones she had taken herself. She drew illustrations and graphs by hand, then labeled her artwork with sticky little letters that looked like type. "I pressed those letters down so hard," she laughs today, "that I got all sweaty. The letters stuck to me, not the paper."

As the deadline loomed nearer and nearer, Mimi brought a sleeping bag into the biology building and worked almost around the clock. "If I didn't hand in the dissertation before the deadline," she explains, "I'd have to wait another whole semester to get my degree."

Finally, with just hours to spare, Mimi finished the dissertation and turned it in. *Hooray!* she thought to herself. *Now I can finally get some sleep!*

A weary Mimi Koehl walked to the parking lot, her Ph.D. work finished at long last. Upon reaching her car, she was greeted by a parking ticket tucked beneath the windshield wiper. "I had parked the car legally," says Mimi with a smile, "but my campus parking permit expired on the same day that my dissertation was due."

As she did her **experiments**,

Mimi learned

some **surprising** things

about sea anemones.

7

More Riddles to Solve

The shipyard workers in Seattle, Washington, didn't know what to make of Mimi when she lugged an outboard propeller into their shop one day in 1977. She politely asked, "Can you mill down the size of these propeller blades, please?"

"Hey, lady," the workers were tempted to tell her, "we handle oceangoing freighters, not dinky powerboats. Besides, these propellers are really small already."

Mimi sensed their confusion. "I don't need it for a boat," she explained. "I'm building a flow tank."

She tried to describe the design. "I told them the tank looks like a big donut filled with water," she recalls. "The propeller needed to fit inside the donut so it could stir the water in the tank. If they milled down the blades, I could use the propeller to control the water speeds inside the donut."

The men nodded their heads as if they understood. But Mimi soon found otherwise. When she returned to pick up the propeller, one of the guys in the shop said, "Oh yeah—you're the lady building the donut mixer."

"They had picked up on two words: 'stirring' and 'donut,'" Mimi laughs. "They thought I was building a machine to stir up donut batter."

Whether she's in the field *(opposite)* or in the lab, Mimi relies on creativity and ingenuity to get her work done. That was certainly the case in Seattle, Washington *(above)*, where Mimi asked local shipyard workers to help her fashion a propeller that would fit the flow tank she was building.

~ *Do-It-Yourself Science*

Mimi discovered early that she couldn't just walk into a scientific supply store and buy equipment. Her research often requires unusual gear. "I've had to invent and build a lot of devices as I go along," Mimi explains. "Those carpentry lessons with my father have come in quite handy."

Mimi got additional workshop training at Duke University in the mid-1970s. One of her mentors, Steve Vogel, made his own equipment, too. He showed Mimi how to use a lathe—a machine in which a piece of wood, metal, or plastic is held and turned while being shaped by a cutting tool. The lathe lets Mimi make parts for the equipment and mechanical models she builds for her experiments.

Students in Mimi's lab build a flow tank, just like she did at Friday Harbor. But, unlike these students, Mimi had worked alone.

While at Duke, Mimi was able to use Steve Vogel's flow tank. After graduating in 1976, however, she needed to build a tank of her own. She wanted to continue her experiments in fluid dynamics—the study of how liquids and gases move. She wanted to keep on linking fluid mechanics to the design of living things.

~ *"I Had to Know the Answer!"*

Mimi built her first flow tank in 1977 at the Friday Harbor marine labs, which didn't have a tank at the time. She had moved there in the fall of 1976 as a postdoctoral student, or postdoc. As Mimi points out, "Just because you have your Ph.D. doesn't mean you're done learning how to be a scientist. You study with another scientist who knows something different than your Ph.D. professor."

Mimi had won a fellowship to work as a postdoc at the University of York in England. But she put off the trip so she could return to

An aerial shot shows the jutting dock and compact labs of Friday Harbor, which is home to many diverse marine algae and marine invertebrates.

Friday Harbor. "I wanted to test the theory that Dan Rubenstein and I had worked on at Duke. In our paper, we proposed a mathematical model about how marine animals catch small food particles from the water. If we were right, the speed of the water flowing past calm-water sea anemones—the tall, white ones with big fluffy crowns of tentacles—would influence how good they were at catching microscopic animals from the moving fluid. I had to know the answer!"

Mimi headed west to Friday Harbor with neither a job nor a big grant to pay her way. Her only funds were the savings she'd scraped together as a graduate student and a small fellowship from a group called Graduate Women in Science.

"I lived off this money for an entire year," Mimi says. "I stayed in a one-room apartment, ate mostly rice, and wore threadbare clothes. I used most of my money to buy research supplies (including an underwater camera) and to build a flow tank. That's how badly I wanted to do those experiments."

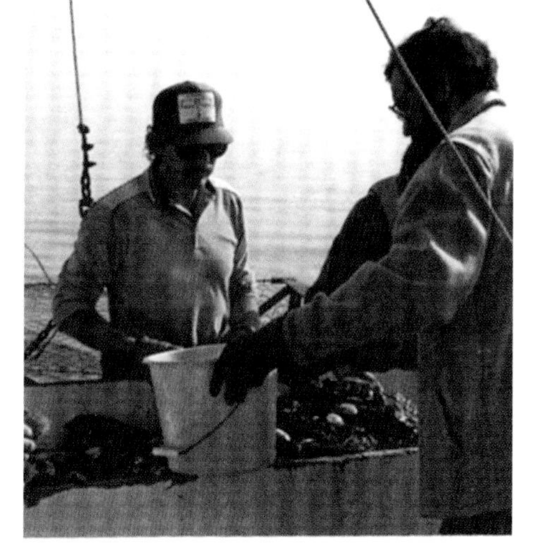

Richard Strathmann *(right)* and a colleague sort through sea life aboard a Friday Harbor research vessel. Asked to explain his passion for studying marine larvae, Strathmann once replied, "Because they are beautiful and I want to understand their shape and form."

Mimi worked with a scientist named Richard Strathmann, a professor at Friday Harbor marine labs. He studies the larvae of marine invertebrates. These larvae hatch from the eggs of many types of sea creatures. For bottom-dwelling marine animals, larvae provide a way to colonize distant habitats. Ocean currents carry the larvae to new sites, where they settle on the seafloor and metamorphose, or change, into their bottom-dwelling forms.

Richard Strathmann wanted to know how the microscopic larvae feed as they drift along on ocean currents. Most of these larvae eventually find a home, but many kinds need to eat along the way. The question was: How do they catch food?

Since both Strathmann and Mimi were interested in how animals catch particles, he agreed to sponsor her as a postdoc. He even found space where Mimi could set up her flow tank.

As she did her experiments, Mimi learned some surprising things about sea anemones. When the water flowed more rapidly, the anemones bent closer to the bottom, where the water flowed more slowly. Not only that, but fast-flowing water pushed their tentacles closer together, making the mesh in their particle-catching nets finer as the currents sped up. The water speed was changing the shape of their food-catching filters.

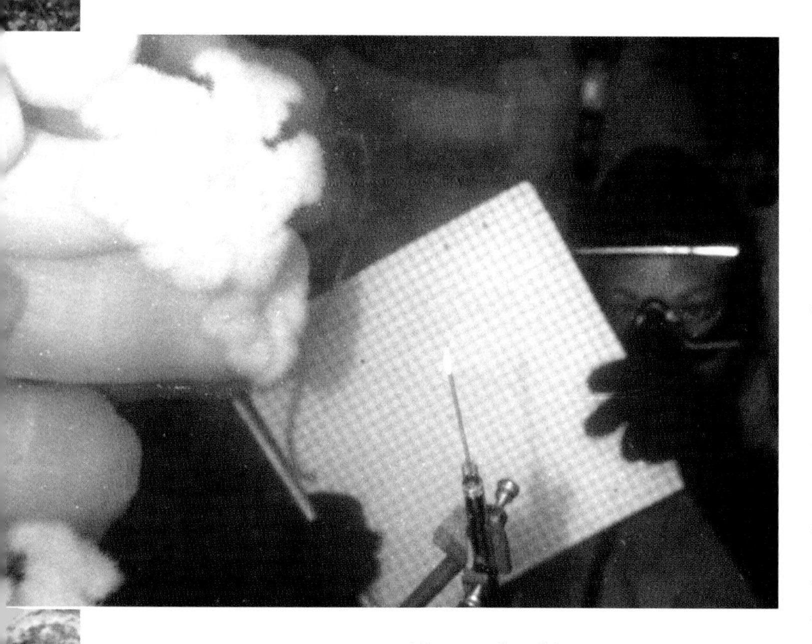

Peering through her scuba mask, Mimi uses a grid square to measure the distance between a current-measuring flow-probe and a cluster of sea anemones growing on a rock wall.

Mimi and Dan had started off thinking that the food-catching filters in some marine animals worked liked furnace filters. But Mimi wound up discovering something else. Unlike furnace filters, the filters in sea anemones change shape, allowing the animals to keep catching food as current speeds change with the tides.

~ *Pounding the Pavement*

"No one can be a postdoc forever," muses Mimi. "You need to find a job, which can be hard. Hundreds of people apply for a single teaching post at a university." Because she worked in an unusual field, she tried to sell herself as a marine biologist or an invertebrate zoologist—a scientist who studies animals that are not fish, amphibians, reptiles, birds, or mammals.

Mimi got called for several interviews. "It was nerve-racking," she recalls. "You had to give a talk in front of all the faculty and students in a department. You also had to meet with each professor, who wanted to figure out how good you were as a scientist."

"My work still mattered, but my illness made me appreciate how wonderful it is to be able to eat an apple or to feel the sun on my face."

At one interview, Mimi got a terrible stomachache. She tried to ignore the pain. "I thought it was just a bad case of nerves, but it got worse and worse."

~ A Gut-Wrenching Experience

Mimi dragged herself home to Friday Harbor. She tried to sleep, but the stomach pain doubled her over. With the help of a student next door, Mimi went to a clinic on the island. The doctor quickly drew some blood for testing.

Mimi's white blood cells—those that fight infection—were alarmingly high. The doctor ordered her airlifted to a hospital on the mainland, where surgeons discovered she had not just appendicitis but peritonitis—an infection that causes inflammation of the abdominal lining. The double whammy nearly killed her.

Mimi underwent two surgeries and spent a month in the hospital. "It really knocked the wind out of me," she says. "But I began to see life as the true gift it is. Before then, I had been wrapped up in getting a job and succeeding with my experiments. My work still mattered, but my illness made me appreciate how wonderful it is to be able to eat an apple or to feel the sun on my face."

Mimi's recovery was agonizing. She had lost 30 pounds and could barely keep down food. Just to walk across the campus at Friday Harbor was a struggle. As Mimi got stronger, she also got good news: She had received not one job offer, but two. Mimi decided to accept a position as an assistant professor at Brown University in Providence, Rhode Island. So in the fall of 1977 she packed her bags and headed to York, England.

Had she boarded the wrong plane? Hardly. The University of York in northern England had held a place for Mimi as a postdoc, and now she decided to fill it. Brown was willing to wait. "It was better for the university," explains Mimi. "I would learn to be a better scientist by having another year of postdoctoral study."

~ *Nature's Needles*

In York, Mimi stepped into a walled medieval city, complete with stone towers and the largest Gothic cathedral in northern Europe. She expected the university to look the same. Instead she found an ultramodern campus filled with high-tech buildings.

Mimi had come to York to work with John Currey, an English scientist doing groundbreaking research in biomechanics. She had met him several years earlier, when he journeyed to Duke to visit Steve Wainwright.

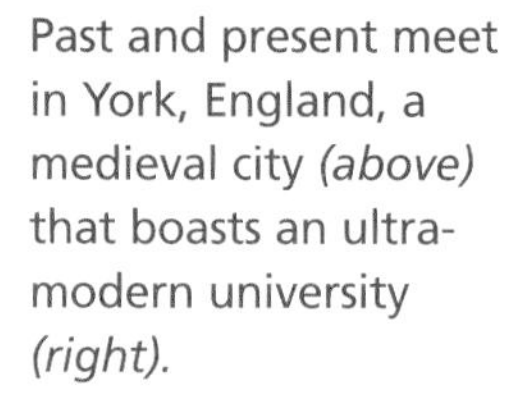

Past and present meet in York, England, a medieval city *(above)* that boasts an ultra-modern university *(right).*

So far, Mimi had focused on the "architectural features"—that is, the structural design—of squishy things in nature, such as anemones and kelp. Now she wanted to look at the hard stuff in nature. John Currey was just the man to help her do that. He was studying the mechanical designs of bones and seashells.

Professor John D. Currey, Mimi's mentor in England, explores the nitty-gritty mechanics of how bones and shells crack and break.

Mimi didn't give up entirely on squishy life-forms. She looked at special types of squishy creatures with hard, needlelike particles of glass or chalk embedded in their soft bodies. These particles, called spicules, can be found in invertebrate animals such as sponges, some types of corals, and sea cucumbers. Although invertebrates lack bony skeletons, the spicules act like a skeleton of sorts.

"Spicules come in all shapes and sizes," Mimi says. "In some animals, the spicules are close together. In others, they're far apart. I began to wonder how these factors—the size, shape, and spacing of spicules—might affect the animal's mechanical behavior."

Mimi intended to find out. She started by comparing creatures with different types of spicules. But different animals have different connective tissues between their spicules. "I couldn't tell whether a certain behavior resulted from the spicules, the tissue between them, or a mix of the two. To tease apart the answer, I went back to building models."

~ *Cooking Up Answers*

Mimi didn't need to model the spicules themselves. She simply cooked, bleached, and burned dead sponges and soft corals until nothing was left except their spicules.

But Mimi needed to pick some kind of squishy material that could stand in for connective tissue. That's because she wanted to focus on differences in the spicules, not differences in the tissue. So the tissue had to be the same in every experiment. Even so, she had to find a material that stretched the same way real connective tissue does.

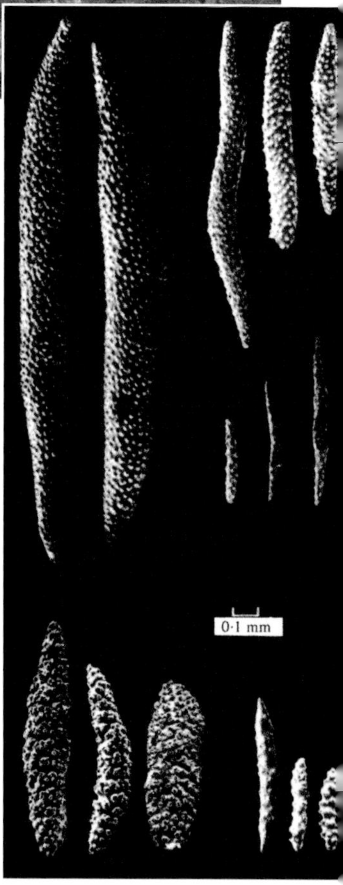

Many beautiful, squishy sea creatures get their shape from sharp spicules. Mimi studied the spicules in pink soft coral (shown at right in the photo above). Detailed views of spicules from a variety of creatures *(inset)* were captured using a scanning electron microscope.

"I finally settled on these dense, rubbery blocks of flavored gelatin I found in a local grocery store," Mimi explains. "They melt when you heat them up. Then you can pour them into molds of different shapes. The flavor didn't matter, so I chose my favorite—raspberry.

"I ended up making molded-gelatin salads. But instead of stirring in raisins and marshmallows, as my mother did, I put in the spicules of marine invertebrates. I added different types of spicules to each mold. I also tested as few or as many spicules as I wanted."

Mimi tested her gelatin tissues the same way she had tested sea-anemone tissues. She measured their stretchiness with an Instron—a special machine that pulled tissue samples and measured how stiff they were.

When "Prof. Currey," as people called him, got a whiff of Mimi's methods, he took to walking into the lab whenever she cooked up some new tissue. "Good God, woman!" he would cry out, pretending to be offended. "You've got the place smelling of raspberries again!"

Amid the banter, Mimi managed to uncover some interesting facts about spiculated skeletons. "I found out one thing you might expect," she explains. "If you put a whole lot of spicules in a tissue, it's much stiffer than if you put in just a few.

"But I also discovered something you might not expect to find. I had thought that great big spicules would make a tissue stiffer than tiny little ones, but that's not true. A whole lot of small spicules reinforce tissue much better than a few really big ones."

It's just one of the things that makes science so captivating. You never know where a raspberry-flavored model will lead you.

~ *Why Not?*

In the fall of 1978 Mimi took her new knowledge to Brown University, where she worked as an assistant professor in the Division of Biology and Medicine. "I had never worked so hard in my life," she says. "I had to create lectures from scratch, lead field trips, design lab exercises for my courses, and advise students. I also had to continue my research, write papers about my discoveries for publication, read the latest articles on biomechanics, and give talks at scientific meetings. I was beat!"

One day an ad in a science magazine caught her eye. The University of California at Berkeley had an opening in its zoology department. "That was where Steve Wainwright got his Ph.D.," recalls Mimi, "so I had already heard many wonderful stories about Berkeley. Plus it's a huge university with lots of scientists doing all kinds of cutting-edge research. When I needed to learn something in another field, as scientists often do, I knew there'd be some expert at Berkeley to get me started."

So Mimi bought some stamps and mailed off her application. It proved to be a wise investment. Berkeley invited Mimi to California for an interview—an event that opened the door to her work today.

Mimi, the only

woman scientist

in the program, comes into view.

8

TESTING TIMES

One by one, mechanical creatures—lion, tiger, bear, rooster, pheasant—lumber across the television screen. An announcer's voice interrupts the parade of toy animals. It tells viewers, "This program is about the most complicated machines on Earth—living creatures."

The camera switches to Steve Wainwright, who introduces viewers to the new science of natural engineering. He's the first of several leading biomechanists to appear in a special broadcast by *NOVA*, the popular TV science series. Each scientist attempts to answer two basic questions: How does an animal work when it does something mechanical, such as fly or swim? And what body features—shape, size, tissue materials, and so on—affect how well a creature stands up to nature's forces?

Lights, camera, action! Opposite and above, a film crew for the *NOVA* science documentary "Living Machines" gives Mimi a few pointers about the microphone at Friday Harbor in 1979. When it came to strapping on scuba gear and plunging into Puget Sound, however, Mimi needed no coaching.

Besides Steve Wainwright, two of Mimi's other mentors, John Currey and Steve Vogel, talk about their work. Midway into the show, the camera zooms in on the Friday Harbor area. Mimi, the only woman scientist in the program, comes into view. She's been chosen to discuss her work with sea anemones. The camera crew catches her diving into the frigid waters of the Puget Sound.

Admitting his bias, Steve Wainwright declares, "Mimi's the star of the whole program."

The documentary aired in February 1980. Mimi may have appeared with "giants" in her field, but she wasn't living like much of a star. Even though she had come through her 1979 job interview at Berkeley with flying colors, Mimi lived as a struggling assistant professor of zoology. She rented a two-room basement apartment, with pipes hanging from the ceiling and a shower drain that belched fumes—and worse—whenever the public sewer lines backed up.

"I was earning a salary," explains Mimi, "but scientists must pay for their own research supplies and equipment. A university expects us to win grants from the Government or a foundation to pay for our work. We write long proposals, explaining why a research question is important and detailing the experiments needed to answer it. Yet only a fraction of proposals written by scientists ever get funded. I hadn't won any big grants yet, so I paid for my own research. The first few years were really tough."

~ *The Coffee-Break Grind*

Mimi now faced a make-or-break situation in her career: She needed to prove herself to earn tenure, or academic job security. Achieving that status gives professors the freedom to openly explore all kinds of ideas without fear of losing their jobs for political reasons. "As you can imagine," explains Mimi, "no college or university offers job security the minute you start work. You go through a proving period, called pre-tenure, in which you must demonstrate three things: an ability to teach, achievement of cutting-edge research, and service to your university and profession."

"I hadn't won any big grants yet, so I paid for my own research. The first few years were really tough."

These criteria add up to a lot of work. Mimi faced the same crushing workload that she had endured at Brown—plus she taught graduate students. As Mimi staggered under these tasks, she kept tripping over a stumbling block: winning a major grant. Not only was Mimi facing tough competition for a limited number of grants, but also she was working in an unfamiliar field.

Each academic quarter during the school year, the chairman of Berkeley's zoology department invited Mimi out for coffee. He would ask her point-blank, "How many papers have you written? Have you won a grant?"

Mimi had published papers, but she kept getting turned down for grants. So over each cup of coffee she had to admit, "No, I haven't won a grant yet."

To this the chairman always replied, "You'll never get tenure if you don't get a grant."

"Those talks with the chairman made me a nervous wreck," recalls Mimi, "but they also kept me going. Every time a proposal got turned down, I redid it and sent it back out."

Mimi stands out from a crowd of male scientists at a symposium in the early 1980s. To encourage colleagues to focus on her work, not her looks, Mimi deliberately sported loose, flowing outfits like this one. That didn't stop Steve Wainwright *(back row, second from left)* from offering his candid view: "You look like a striped barnacle."

~ *Sister Power*

Pre-tenure professors, in many ways, live under a microscope. Mimi felt people looked at her more closely because so few women were science professors at the time. She remembered a woman professor from Berkeley, cell biologist Beth Burnside, who had given a guest lecture at Duke University when Mimi was a student there. So when Mimi started working at Berkeley, she contacted Beth.

"Beth had just earned tenure," Mimi remembers. "She was really busy and didn't know me at all, yet she took me under her wing and helped me sort through the tenure process. Beth didn't have to do this, so I'm doubly grateful. I use her as a role model whenever I meet young women scientists today."

Mimi had another mentor at Berkeley, George Oster. "He was interested in using physics to study biology," says Mimi, "so I discussed new ideas with him—often as we jogged or ran."

Beth and George introduced Mimi to other scientists. Through these contacts, Mimi learned about a special problem-solving group made up of young women scientists like her. She joined the group in 1981 and still belongs to it today.

Two members recall meeting Mimi. "None of us had grown into our confident selves yet," says Suzanne McKee. "We were at the start of careers and faced common fears about our success." Adds Ellen Daniell, "Some of my earliest memories of Mimi revolve around her great sense of humor. Her ability to laugh carried her through some tough situations."

Here's how the group works: At meetings every other Thursday evening, each person describes a problem she is facing in her career. Everyone else listens without interrupting. Then the women discuss the problem.

"After you've worked on your problem," Mimi continues, "you have to make a contract with the group about how you'll fix it. You report on your progress, or make a new contract, at the next meeting."

More than anything else, perhaps, the group is a real confidence builder. "It's tough working in fields with so few women," Mimi says. "So we allow time for something we call 'strokes.' We pay a person a compliment, such as, 'That's a great paper you wrote.' Now, here's the hard part: The person has to take credit for her achievement by agreeing with the compliment, 'Yes, I did write a great paper!' "

Mimi *(right)* and hiking partner Ellen Daniell prepare to set out for a favorite spot—Desolation Wilderness, high in the Sierra Mountains. In her left hand, Mimi clutches a topographic map, an essential tool for backpacking in the wilderness.

~ Don't Shave Those Legs!

Many of Mimi's rejected grant proposals involved copepods—tiny, shrimplike sea animals about the size of sesame seeds. The rejections surprised Mimi, for copepods are vital to life on Earth. They eat single-celled algae—the base of the marine food chain—and then are eaten in turn by bigger animals. In fact, copepods are the most numerous animals in the oceans.

Because copepod feeding plays such a big role in marine ecology, Mimi wanted to figure out how they catch their food. But oceanographers thought they knew the answer: Copepods strain algae from the water using a pair of special sieve-like legs.

As Mimi explains, "Copepods have a row of hairs on each of their food-catching legs, and covering the hairs are tiny bristles. Each leg looks like a miniature fishing net. Everybody assumed that the copepod species with coarse nets strained big algal cells out of the water, whereas the ones with fine nets strained out little algal cells. But no one had actually ever tested this."

> "Some of my earliest memories of Mimi revolve around her great sense of humor. Her ability to laugh carried her through some tough situations."

What got Mimi interested in examining tiny hairy legs? A former classmate from Duke, oceanographer Tim Cowles, had discovered that copepods are switch-hitters when it comes to eating. They can nab big or little algal cells, depending on which happens to be most abundant in the neighborhood. But if copepods simply sieve their food from the ocean, how do they switch back and forth between netting big and little algae? It was a mystery.

Tim took his copepod-feeding mystery to Mimi and asked, "Want to help me solve it?"

"It sounded like fun," Mimi recalls, "so I said, 'Sure!'"

~ *Beast on a Leash*

Tackling the copepod mystery demanded some creative techniques. Copepod legs are too small to observe with the naked eye. They must be viewed through a microscope. Put a copepod in a drop of water on a slide, however, and you are no longer viewing it in real-world conditions. "It's like trying to figure out how people swim by watching them thrash about in a wading pool," Mimi laughs. "Instead, we needed to watch a copepod through a microscope as it swam around in an aquarium."

Another scientist, Rudi Strickler, had built a fancy system of camera lenses that made it possible to do just that. So Mimi and

Tim traveled to Rudi's lab in Canada to look at copepods through his lenses.

Copepods flap their feeding legs so rapidly they look like a blur. The two scientists solved this by filming the copepods with a special camera that captured 500 frames per second. Then they studied the leg motions by reviewing the movies frame by frame.

The moviemaking scientists struggled with the stars of their films. "It's hard to persuade copepods to catch food at the precise spot where your camera is pointed," says Mimi. "So we put them on leashes! How on earth do you leash a microscopic animal?"

Rudi suggested a novel solution. "We dipped the tip of a human hair in Crazy Glue," Mimi explains, "then stuck it to the back of the copepod. (Copepods have hard shells, so it didn't hurt them.) Then we guided our leashed copepod to the spot in the aquarium where our camera was focused. Because our hands were too clumsy to guide a microscopic beast on a leash, we attached the hair to a precise instrument that let us move the leash by turning little dials."

A single frame of a high-speed movie *(top)* catches a copepod in the act of flinging its food-catching legs. Using a scanning electron microscope to shoot close-ups of the animal *(above)*, Mimi could measure how the hairs and bristles were arranged on each species.

Mimi and Tim also needed to see how the water moved around when a copepod flapped its feeding legs. Mimi mixed some seawater with food coloring. Then she fashioned a microscopic glass nozzle that could release a fine stream of colored water near the copepod. This allowed her to watch the colored water move around as the copepod fed on big or little algal cells.

~ Dining Instructions: Fling, Wiggle, Slurp

The films' debut raised a lot of eyebrows. By going through the movies frame by frame, Mimi discovered that water did not flow through the nets on the copepods' feeding legs. "They weren't straining particles out of the water at all!" exclaims Mimi. "They did something entirely different."

Here's how a hungry copepod stuffs itself: When it gets near a big algal cell, it flings apart its hairy feeding legs and draws water into the space between them. The algal cell gets carried along with the water toward the copepod's mouth, which lies between its feeding legs. Then the animal moves other little appendages that sweep parcels of algae-filled water into its mouth.

Mimi made more movies of copepods with Gustav-Adolf Paffenhöfer, who had lenses and a camera like Rudi's at the Skidaway Institute of Oceanography. They discovered that when the water near a copepod contains only little algal cells, a copepod switches its behavior. It rapidly wiggles its feeding legs up and down, pumping a steady stream of water and tiny algal cells toward the animal's mouth. It then slurps them up.

Excited by the experiments, Mimi started writing and speaking about what she and her co-workers had found. She presented just the facts, but they popped a lot of bubbles. At the first scientific conference she addressed, one copepod expert piped up, "Physics might apply to your copepods, but it doesn't apply to my copepods."

Scientists reviewing grant proposals felt differently. They now considered Mimi's work so groundbreaking that they finally awarded her a grant.

~ The Terror of Tenure

In 1982 Mimi got a huge surprise. She was on sabbatical—an academic leave of absence to do research. That August she returned to Berkeley

to go on a backpacking trip with Zack Powell, her future husband. When Mimi checked her office mailbox, she found a note from the department chairman. It read, "We're putting you up for tenure early."

Mimi had planned to climb through the Sierra mountains with Zack. Now she had to climb through a mountain of work. Mimi had so little confidence in herself, she thought, *Oh, no! I've done such a bad job that they want to get rid of me early!*

Zack Powell *(above)*, is pictured at about the time he and Mimi started dating. Below, Mimi pokes her head above a boulder in the mountain pass where she and Zack went hiking to calm her "tenure terrors" in the autumn of 1982.

Her fears were deepened by the fact that Berkeley had not given tenure to her close friend Ellen Daniell. Ellen had brought the issue to her problem-solving group, which helped her plot out a new career. Mimi now turned to the group, too.

"When Mimi got called early," Ellen admits, "it was tough for me. I was happy for Mimi but sad for myself. But we talked about it in our group and stayed good friends. In fact, Mimi told me that she learned about the rigors of tenure from my experience."

University policy said Mimi was entitled to a full six years in which to prove herself. "I went to my department chairman to ask

for more time to get ready," Mimi explains. "But when I got to his office, I broke down in tears. I was so embarrassed."

The chairman listened to Mimi speak. When she finished, he said, "Don't you think the department understands how the university works better than you?" He went on to say that Mimi had done more to merit tenure in three years than the six years normally given to people.

"It turns out," Mimi smiles today, "that I worked twice as hard as was required to keep the job that I was terrified of losing."

After handing in her tenure papers, Mimi and Zack headed for the Sierras—only to get a flat tire en route. They reached the mountains late in the afternoon, then trudged through deep snow for hours to reach a high pass just at sunset.

"We saw the most amazing red sky," Mimi remembers, "and gorgeous clouds changing color. I was eager to distance myself from the early tenure call. So for the next eight days, I backpacked away my stress."

The story has a happy footnote. After reviewing Mimi's credentials for nearly a year, Berkeley granted her tenure as a professor.

Mimi and Sharon

learned that ***flying frogs***

worked like jet fighters.

9

Flights of Fancy

Like a sluggish bee on a chilly day, a five-inch-long insect moves slowly in the warm morning sun. The cold-blooded creature can't flap its wings and lift off the ground until its body temperature heats up.

But wait a second—something's wrong here: This insect doesn't have wings. It's got thin, stubby little body flaps. It's prehistoric times, and insects haven't yet developed wings.

Welcome to Mimi Koehl's lab, circa 1984, where she and a postdoctoral student in biology named Joel Kingsolver have zoomed back some 350 to 400 million years ago. The scientists are trying to unlock a mystery: How did these stubs evolve into wings?

Some research projects are just plain fun, such as the one in which Mimi figured out how a tree frog *(opposite)* glides. Above, she kids around with colleagues at a science meeting in 1987.

~ *A Mystery of History*

The mystery had grabbed Mimi's imagination. "We know from fossils," she explains, "that the most primitive insects had no wings. In later—but still ancient—fossils, we see insects with stubby flaps on their bodies. In even later fossils, and on modern insects, we find long, lovely wings capable of flapping flight. Joel and I wanted to figure out how stubby flaps might have evolved into long wings."

To begin solving the riddle, Mimi and Joel thought about the process of natural selection. In any population, individuals look slightly different from one another. That means some early insects probably had longer body flaps than others. Suppose, over the course of many generations, that more long-flapped insects survived to produce young than did short-flapped insects. The insect population would eventually be made up of insects with longer flaps. The two scientists wondered: What could the insects with slightly longer flaps do better than the ones with shorter flaps?

~ *The Flap Over Wings*

All sorts of scientists had already pondered this question. One group thought the pre-wing flaps might have served as parachutes to slow falls. A second group suggested that the insects used the flaps to glide from place to place. A third group guessed falling insects used the flaps to steer, helping them land right-side-up. Yet a fourth group threw out all these theories. It said that the flaps acted as solar panels, not flight gear.

To build a model of a flying insect in the present, Mimi had to study how flight evolved in the past. Shown below is one of the fossilized insect wings she examined. It measures about 15 centimeters, or 5.91 inches, in length.

"Without a time machine," Mimi laughs, "you can't go back to test any of these theories. So people simply argued about which of them was right. Joel and I thought all the ideas sounded reasonable, so we decided to test them with experiments. That's how science works: You state a theory as a hypothesis, then design experiments that try to prove it wrong. Unless you follow that process, you're debating beliefs, not doing science."

In designing experiments, however, Mimi and Joel faced a problem: These primitive flap-winged insects were extinct. How could the scientists test them? With models, of course!

~ *Picturing the Past*

Mimi and Joel pored over pictures of insect fossils and looked at real ones in museums. The two scientists learned that insects with flap-like wings came in all shapes and sizes. They picked the two most common body shapes: a wide-bodied form and a slim, cylindrical form. Then they went to work building models—small, medium, and large—of each one.

"It was tricky picking the right materials," recalls Mimi. They had to make the insects out of a special epoxy, or glue, that would conduct heat the same way real insects do. The wings were an extra challenge. Needing to create thin, lightweight membranes, they wound up designing delicate wire outlines of wings and dipping them in a plastic goop used to make fake flowers.

Mimi and Joel also needed to re-create the world from the point of view of a prehistoric insect. "When you ride a bike, you move through the air," explains Mimi, "but it feels like wind is blowing on you. To an insect falling through the air, it would have felt like wind blowing past its body. We used a wind tunnel to move air past our models just like it would flow past a falling or gliding insect."

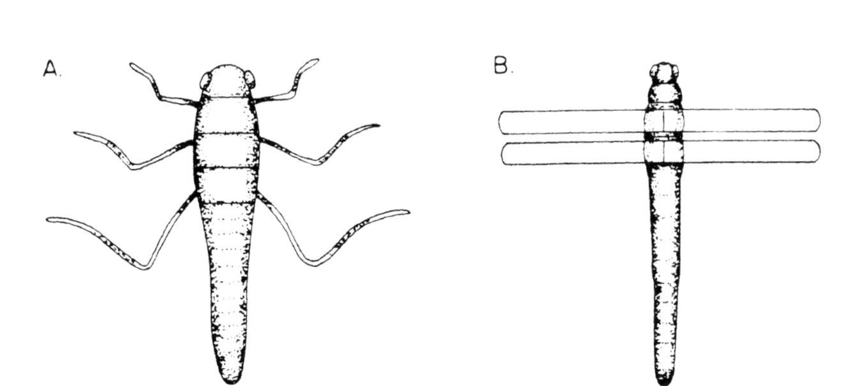

Mimi drew these insect-model diagrams for a scientific article she published. Diagrams A and B are top views of the two basic model shapes that she and Joel Kingsolver used to re-create Paleozoic insect flight. Model B wears a pair of long wings.

~ Time Travelers

The stage was set for a journey into the past, and Mimi and Joel were the perfect time travelers for the trip. From his study of butterflies, Joel knew how insects use their wings to regulate temperature. He designed experiments to test the "solar panel" theory. Mimi, an expert in fluid dynamics, explored how flaps might affect the way insects moved through air. This meant applying the same physical principles that an engineer uses to study airplane flight.

Joel Kingsolver *(right)* and Mimi worked together to build realistic models of prehistoric insects like the one below. Then they tested the models in the controlled environment of a wind tunnel *(below right)* and recorded the results.

Mimi and Joel wired their models to electrical instruments that measured how much they heated up and how well they parachuted, glided, or steered. Would models with longer flaps do any of these things better than models with shorter flaps?

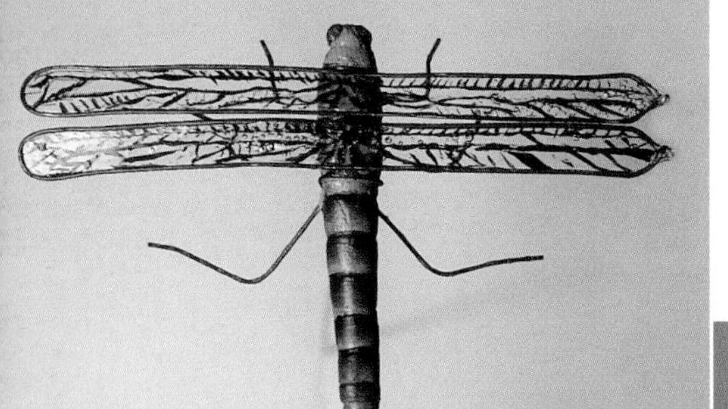

"We made all kinds of wings," says Mimi, "from short stubs to elegant long ones. We'd dress up a model with one set of wings and take it for a spin in the wind tunnel. Then we'd change its wings, or add even more wings, and take it on another test flight. Next we would remove all the wings and test it again."

When Mimi and Joel analyzed their data, they learned that stubby flaps made great solar panels. When they lengthened these flaps, the wings did an even better job of warming up the insect. So the solar-panel hypothesis could explain how short stubs evolved into longer ones.

It was a different story when Mimi and Joel made medium-sized wings even longer. "We measured no warming effect," notes Mimi. "The heat absorbed by the tip of a long wing is lost to the surrounding air before it can travel up the wing to the insect's body. When we realized the solar-panel hypothesis could not explain why medium-sized flaps evolved into very long wings, we threw it out."

> The stage was set for a journey into the past, and Mimi and Joel were the perfect time travelers for the trip.

But what about those theories of how flaps affected early-insect flight? "The ancient short-flap insects, we discovered, were really lousy at parachuting, gliding, and steering," says Mimi. "But when we stuck medium-sized flaps on our models, the insects got better at those functions. When we made the flaps longer still, the insects did even better on their test flights."

~ *Drawing Fire*

Mimi and Joel were not proposing some radical new theory. They had merely figured out a new way to test existing theories to see which ones worked and which ones didn't. But their work with prehistoric flight captured the public imagination. Reporters showed up to interview the two scientists. Newspapers quoted them; radios broadcast stories about their work.

Some scientists who had been arguing about the origins of insect flight now started questioning Mimi and Joel's findings. "Joel and I were both newcomers to the field of insect flight," Mimi explains. "We had fun doing the experiments, but it stopped when people started attacking us."

For the time being, Mimi decided to steer clear of hotly debated topics such as the origins of flight. Meanwhile, other exciting research projects and honors awaited her.

~ Honor at Gettysburg

One honor came in 1985 when Gettysburg College awarded Mimi its Young Alumni Achievement Award. The school invited the former art major to give a lecture to its biology department.

Mimi's mother died shortly before Mimi got the news. Meanwhile, Mimi's dad suffered from heart problems and didn't have much longer to live either. Zack offered to take him to Mimi's talk, but he didn't want to go. Zack, who had recently lost his own father, told Mimi's dad that he deeply regretted that his father, a Ph.D. chemist, never saw him teach. He then said, "We don't know if you'll get another opportunity like this. Mimi has paid you a great compliment. She followed in your footsteps and became a science professor. You should see how very special she is."

Bob Koehl and sister Mimi pay a windblown visit to San Francisco's Golden Gate Bridge in 1984. "When our father died in 1987," says Bob, "I took over the job of button-busting."

Mimi's dad said little. He walked into the bedroom to change for the trip. After the lecture, he turned to Zack and said, "She really is something special, isn't she?"

According to Mimi's brother Bob, their dad came home from Gettysburg "busting his buttons with pride." But he never told Mimi directly. The news came secondhand, which as Zack says, "was very sad for both Mimi and her dad." He then adds, "But Mimi knows the story now, and I'm pleased that I can tell it."

Mimi's pleased, too. You can see by the smile that crosses her face when she hears Zack recall the day when her father listened to her lecture. George Koehl died in 1987. Mimi was still only in her thirties.

~ Pink Flying Frogs

In the late 1980s, Mimi fielded a question about a different sort of airborne creature. Sharon Emerson, a biologist at the University of Utah, asked Mimi to help her understand the mechanics of

tropical flying frogs—the kind that glide down from treetops to lay eggs in pools of water on the rain forest floor. Sharon knew why the frogs jump out of trees, but she didn't know how their weird bodies affected their flight.

Mimi and Sharon Emerson *(foreground)* show off their frog models. With the help of a student assistant, the women conducted test flights in the handmade wind tunnel at right. The diagrams above show the two basic plastic models that Mimi and Sharon tested. Both types had detachable hands, feet, and flaps.

Mimi agreed to help Sharon figure this out. She started by looking at the design of flying frogs. Unlike chubby regular frogs, flying frogs have slim, airplane-shaped bodies and enormous webbed hands and feet. Flying frogs move differently, too. When they leap from trees, they bend their knees and spread out their huge webbed feet. Regular frogs, which have small feet, push their legs straight out. *What effect,* she wondered, *did features such as bent knees and fancy feet have on the frogs' ability to fly?*

Mimi turned to model-making to learn the answer. Sharon supplied pickled frogs—both the flying and regular varieties. They convinced a dentist to make molds of the frogs and to cast models of them, using the pink plastic he used to make retainers.

Mimi and Sharon added changeable body features. They could put big, flying feet on a fat, nonflying body. They could bend the knees or pull the legs out straight. "We even designed accessories like hats and pocketbooks," laughs Mimi. "But we didn't use those in our experiments."

The frogs would not fit in an insect-sized wind tunnel, so Mimi and Sharon built a new one. They made it from purple posterboard, duct tape, and a high-powered fan.

Nor did the women work only in the lab. Mimi tested the models by pitching pink frogs off of the deck of her hillside house. Sharon, standing down below, called out the movements: If a frog tumbled end over end through the air, for example, Sharon shouted, "Pitch!"

One curious neighbor raised his window, stuck his head out, and asked, "What are you two doing?"

"We're testing the aerodynamic stability of flying frogs," Mimi exclaimed.

"Aha," said the neighbor. "Of course." He quickly ducked back inside.

Flight in a Tunnel

Imagine riding your bike at 5 miles per hour (mph) on a windless day. As you ride, it feels like wind is blowing on your face because you're moving through the air. Now imagine sitting on a bike in front of big fan that's blowing air at 5 mph. You feel the same wind on your face. Even though you're not moving through the air, the air is moving past you, causing the same effect. That's because in both cases, the air is moving at 5 mph relative to you.

Now think about a frog gliding through a rain forest. Winds skirt the tops of tangled branches high in the forest's canopy, with hardly a breeze below. Yet as the frog falls through the air, the air moves relative to the frog's body. It moves at the same speed as the frog, but in the opposite direction. So a frog falling downward feels the wind blowing upward.

To figure out how the body shape and posture of a model frog affect how well it glides or maneuvers, scientists measure the forces that the air exerts on the model as it falls. To do this, they attach the model to a measuring device. But the model can't move if it's attached to the device, so what do they do? They create a wind tunnel! A wind tunnel is basically a fancy fan in a big tube. The fan is controlled to blow air at just the right speed, and devices in the tube let scientists adjust factors like wind turbulence. A free-flying animal in a wind tunnel "flys in place," just like a person on a treadmill runs in place.

If the wind in a wind tunnel blows past the stationary model at a frog's gliding speed, the air flow relative to the model will be the same as it is on a free-gliding frog. If the air flow relative to the model and the gliding frog is the same, then the forces that the air exerts on them will be the same, too.

A real flying frog glides through the air inside the wind tunnel. (It landed safely.)

A short time later, the little girl next door yelled out the same question. This time Mimi replied, "We're playing with pink frogs."

"Nice!" the little girl replied.

"Evidently," Mimi laughs, "that was a much clearer answer."

~ *Steering Clear of Trouble*

Mimi and Sharon learned that flying frogs worked like jet fighters. Think about the design differences between a passenger jet and a fighter jet. Passenger jets are designed for stability: They resist being blown around by wind currents. Yet that very design resists turns by the pilot, making them hard to maneuver. (That's why passenger pilots go around in big circles when they land.) A fighter jet, by contrast, can make tight turns and do aerobatics, but they're highly unstable.

Now look at a tropical rain forest through the eyes of a flying frog. It's packed with trees for you to bonk into—and breeding ponds are small targets to hit on the forest floor. Striving to survive and breed, you might happily give up stability for steering.

The lessons Mimi and Sharon learned from flinging their frogs and from test flights in the lab proved this trade-off in nature. Regular frogs tend to be stable; if they fall from a tree, they will probably land right-side-up. Flying frogs tend to be unstable—their slim bodies, enormous feet, and bent knees make them so. But they excel at the sharp turns that keep them from crashing into trees.

These findings put Mimi in the news again. Just a year earlier, in July of 1990, she had won her MacArthur Fellowship (an honor that Sharon Emerson, too, would receive in 1995). Reporters took a renewed interest in Mimi's projects. Some turned up at her house to watch her toss those pink plastic frogs from her porch.

Mimi showed reporters that serious science has its fun side, too. But as she pitched pink frogs, she did something nobody else had done. She used flying frogs to understand the aerodynamic stability of gliding animals. Yet despite these flights of fancy, Mimi's first love remained the turbulent sea.

The big **bird** waltzed
with a bunch of wedding guests,

including the **bride** and groom.

10

SCENT OF SUCCESS

Here comes F. L. Mingo, and the bird is sneaking up behind Bob Paine to steal a dance with him. Bob doesn't see the fuzzy pink creature as it shuffles toward the dance floor. Bob hates to dance, but Mimi has made it hard for him to say "no."

"It's my wedding day," she reminds him. "You have to dance with me!"

"Oh, all right," Bob grumbles.

F. L. Mingo—a name Bob and Mimi have invented while pulling flamingo jokes on each other—strikes quickly. As Mimi and Bob start dancing, a living, breathing, human-sized bird taps Bob on the shoulder.

Bob's jaw drops open when he sees a black-beaked bird, complete with a feathered scarf, standing behind him. F. L. Mingo lives!

"Once I stopped gawking," Bob admits, "I knew that Mimi had gotten me back for the F. L. Mingo jokes that I had played on her. She designed and sewed the costume herself, then persuaded a crazy scientist friend to wear it."

F. L. Mingo made quite a splash that day. The big bird waltzed with a bunch of wedding guests, including the bride and groom. After a long courtship, that's how Mimi and Zack celebrated their marriage in February 1993.

To Mimi's delight *(opposite)*, F. L. Mingo makes an appearance at her wedding to Zack Powell in 1993. Mimi sewed the fluffy pink costume so she could play a joke on Bob Paine. Mimi and Zack can be pretty silly *(above)*.

~ *Scheduling a Life*

"It was love at first sight," says Suzanne McKee of her two friends, Mimi and Zack. Even so, it took work for the couple to get beyond a few scientific discussions. Zack asked Mimi on a date three times, but she turned him down.

"I was too busy each time," recalls Mimi. "And after three turndowns, I just knew this guy would never ask me out again."

"I can take a hint," laughs Zack.

Mimi laughs, too. "The fact is, I really did want to go out with him. I thought he looked kind of cute, and I knew from our talks about ocean currents that he was really smart. But I was too shy to call him for a date. So I sent a postcard inviting him to a modern dance performance at Berkeley."

All right! After an 11-year engagement, Mimi and Zack are all smiles at the conclusion of their marriage ceremony.

Mimi waited for Zack to call her. Instead, recalls Zack, "I cut big letters from newspaper headlines and sent what looked like a ransom note. I wanted to make a strong impression."

"He succeeded," says Mimi. "When I opened that letter, I burst out laughing. This guy may be a little weird, I thought, but I bet he's a lot of fun."

Mimi and Zack have more in common than a good sense of humor. They both enjoy the outdoors. They're fond of good food. They like football, concerts, and art galleries. And they adore science.

As tenured professors and research scientists, however, Mimi and Zack are often pulled in opposite directions by their professional duties: oceanographic cruises, field trips, science conferences, lecture tours, and more. This makes it tough to plan any sort of personal life—even a wedding.

When the couple finally scheduled the ceremony, scientist friends flew in from all across the country. "It looked like a mini science conference," Mimi jokes. "All these people sat around at our wedding discussing their research. It was great!"

Mimi shares her wedding-day excitement with the two "main men" in her professional life, mentors Steve Wainwright *(left)* and Bob Paine. "Bob dressed up for the occasion," Mimi recalls. "Steve, believe it or not, dressed down!"

"At the time," says Steve Wainwright, who traveled to the wedding from Duke University, "Mimi had lots of projects to talk about. And she did so with passion and joy."

Some of the work that Mimi talked about with great enthusiasm involved little hairy noses, especially lobster noses.

~ Smells Fishy

A lobster's nose knows how to sniff out a meal. Its nose—a pair of small, hairy antennules—plucks odors from the water as the hard-shelled creature travels around the murky ocean floor. A lobster uses its highly developed sense of smell to find food and to recognize other lobsters. (It's how boy meets girl.)

The U.S. Office of Naval Research wanted to sniff out how a lobster's keen sense of smell works. Perhaps this hairy little nose could help engineers design underwater robots able to track down toxic sites—places too dangerous for scuba divers to investigate.

Mimi was one of many scientists who wrote proposals for research funding on how animals catch odors and use them to navigate in the ocean. "I'm not in the business of robotics," Mimi explains. "But I am interested in how organisms function in fluids. I had been studying how animals such as copepods use their hairy legs to catch particles from water. Now I wanted to understand how rows of hairs catch much smaller things—molecules—from the sea. Hairy lobster noses were perfect for this."

~ Making Sense of Scents

To make sense of scents, Mimi looks at a couple of things. "For one, I'm interested in how currents of wind or water distribute odors. I ask where the smell is really strong, and where it's weak."

Mimi also tries to figure out how organisms interact with these patterns of smells in the environment when they sniff. As she puts it, "I'm eager to learn about the very first step in smelling—the act of catching odors by noses, antennules, or whatever. How does a living thing catch odor molecules? Which features of a nose affect how well it catches aromas?"

Mimi leaves it to another group of scientists, known as neurobiologists, to look beyond the nose—that is, to examine how the brain interprets odors. "They're the ones," Mimi explains, "who study how the brain figures out whether an odor means dinner, danger, or romance."

~ Nervy Hairdos

Early in Mimi's career, she had noticed that all kinds of hairy appendages help animals do important things—swim, catch food, take in oxygen. The appendages have many different hairdos, or patterns in which their hairs are arranged. Mimi wondered, *Do some hairdos work better than others when it comes to catching odors?*

To answer this, Mimi needed to compare noses with very different hairdos. She picked three types of crustaceans: mantis shrimp, crabs, and lobsters.

"Watch any of these animals," says Mimi, "and you'll notice two sets of 'feelers' up front on their heads. The little ones, which you can easily see on lobsters in a tank, flick all the time. But they're not feelers at all—they're really smellers covered with hairs. Each hair is chock-full of hundreds of chemosensory neurons—nerves that send a signal to the brain if the right kind of molecules hit them."

Flick, pause. Flick, pause. That's what crustaceans do with their busy antennules. But what odor patterns are they sampling? And how do the antennules succeed in plucking odor molecules from swirling currents?

To find out, Mimi and her students examined crustacean hairdos with a scanning electron microscope. Mantis shrimp, they observed, sport just three nose hairs on each segment of their antennules. Crab hairdos look like toothbrushes. And lobster noses have the fanciest styles of all, with hundreds of hairs arranged in zigzag rows.

Sniffing out food, friends, and enemies, a spiny lobster—the kind studied by Mimi—noses around its habitat: a Florida coral reef. When it detects a promising scent with one of its hairy antennules (circled above), the lobster can follow the odor trail.

Every once in a while, lobsters take a break from flicking and start picking their noses. That may sound gross, but lobsters don't live to smell—they smell to live. Unless they unclog dirt from their nose hairs, odor molecules in the water can't reach the nerves inside those hairs.

~ *Robo-Lobster*

Mimi still didn't understand how odor-carrying water got into the tight spaces around lobster nose hairs. So she got busy building models. As Mimi recalls, "One of our lobsters had shed its shell to grow bigger. We built a mechanical lobster from the discarded shell and put two thin wires where the nose goes. Then we clipped a pair of antennules off a frozen lobster from a seafood supplier and slipped them over the wires like gloves. Finally, we hooked this robot to a computerized motor so we make it flick its antennules."

Why didn't Mimi use a real lobster for this experiment? For one thing, she wanted to try out different hairdos. A mechanical lobster wouldn't mind if Mimi stuck a stranger's nose on it. For another, a "robo-lobster" didn't care if Mimi flicked its antennules at the wrong speed.

Mimi wanted to witness a lobster nose in the act of catching odors. "But just try to persuade a real lobster to flick its nose exactly where a camera setup is pointed!" exclaims Mimi.

Mimi took her robotic lobster to Jeff Koseff, an engineer who had a big flow tank in his lab at California's Stanford University. "The tank could mimic the turbulent currents in nature," says Mimi. "All that swirling stirs up smells. So we set it up to duplicate the conditions where lobsters live."

~ *Flicker of a Sniffer*

Because smells are invisible, Mimi and Jeff used fluorescent dye to mimic an odor plume. They put the mechanical lobster in the flow tank and turned on the current. Like the aromas of a lobster's lunch wafting over the seabed, fluorescent dye oozed from a pump on the floor of the tank.

They shined a thin sheet of laser light through the dye cloud near the lobster's nose hairs. In a darkened room, only the slice of dye hitting the antennule lit up. This allowed them to see what happened when the antennules flicked through the dye, or "odor."

Mimi's mechanical lobster flicks its automated "nose" through a plume of fluorescent "odor." The laser light slicing through the plume allows scientists to see only the part of the plume that the nose encounters as it flicks.

"A lot of people were needed to run these experiments," Mimi notes. "Someone had to manage the flow tank. Someone else had to operate the laser. A third person controlled the lobster. A fourth ran the odor pump, and a fifth activated the high-speed, computerized camera. Jeff's students and my lab assistants all pitched in."

The team effort paid off. Only during the rapid downstroke of the antennule flick, as the crew discovered, do smells enter the tight spaces around a lobster's nose hairs. During the slower upstroke, the odors stay trapped. Then, during the next downstroke, the old smells are swept out and a new sample of water moves in to surround the hairs. In other words, the lobster takes a fresh sniff of water each time it flicks its antennules. Flick, flick means sniff, sniff.

~ *Striped Odors*

What do lobsters make of all these smells? The answer isn't clear yet, but Mimi and her research team found something that might provide a key.

"By observing the path of fluorescent odors in the flow tank," says Mimi, "we discovered that aromas come in stripes, not clouds. An odor plume, as we call it, is made of many fine filaments, or strands, of stinky water swirling around in clean, nonstinky water. So when a lobster flicks its antennules, it catches a certain pattern of filaments. It's kind of like mapping a smell. This odor pattern changes as the lobster gets closer to or farther away from the odor source, such as a tasty meal."

Today Mimi continues to study how lobsters catch odor information swirling in the water around them.

To see how lobsters can figure out where they are in a stinky plume swirling through turbulent waters, Mimi and Jeff have teamed up with a group of neurobiologists. They want to learn if lobsters can recognize different patterns of odor filaments. "Even if lobsters can't read these odor maps," declares Mimi, "perhaps engineers can use our work to build robots that can."

Today Mimi continues to study how lobsters catch odor information swirling in the water around them. But she is checking out another stink as well. Many creatures that inhabit the ocean bottom spew microscopic larvae into the water. Mimi wants to know: How do these larvae use odors to find new homes on the seafloor?

Mimi and Mike placed

scientific instruments

in and above a coral reef

in Kaneohe Bay, Hawaii.

11

That's My Cue!

"Scientists in remote places," Mimi has remarked, "must be prepared to come up with creative solutions in case something goes wrong in the field." Nothing could have been more true than when Mimi and Zack headed off in the early 1990s to do a joint research project in Palau, a chain of coral islands southeast of the Philippines. Something went wrong before they even landed. They touched down in Palau without a single piece of their scientific equipment in the belly of the plane. Everything had been lost.

Mimi snorkels over a coral reef in Hawaii *(opposite),* studying how larvae sniff out new homes and settle down in them. Above, she directs volunteers who are helping her conduct an experiment in the transport of marine larvae.

"All I had was my carry-on bag," Mimi groans. "It contained a bathing suit, a sarong, a muumuu, a camera, some fluorescent dye, and my mask and fins. What kind of experiments can you do when that's all you have?"

"Blobs," answers Zack. "We did blobs."

~ *Blobs from the Blue Lagoon*

Because Mimi and Zack work on different scales, a lot of planning had gone into the trip to Palau. As an oceanographer, Zack works on a large scale, looking at the interaction of physics and biology in ocean ecosystems. As a biomechanist, Mimi works on a small scale, examining how individual organisms interact with fluids. But the couple was willing to search for an in-between scale on which they might work together. "Given our busy schedules," Mimi explains, "we thought we'd probably see a lot more of each other if we did a research project together."

Larvae-producing creatures carpet the lagoons and reefs of Palau, an island chain in the South Pacific. The extreme natural beauty of the place inspired the governments of Japan, the United States, and Palau to open a nonprofit marine park and coral reef center there in 2001.

The two scientists decided to study water flowing across distances of a few inches to half a mile. As they talked about the marine processes that occur at that scale, they posed a puzzling question: How do turbulent water currents mix up and spread the tiny, free-swimming larvae spewed out by bottom-dwelling animals?

Mimi and Zack started studying the transport of larvae along the coast near their California home, but jumped at the chance to join an expedition to Palau in 1991. Then the luggage disaster hit. How would they conduct research without scientific equipment?

Mimi hit on an idea. The fluorescent dye in her carry-on bag would let them see how the water moved in the coral lagoons. Mimi explains, "We released blobs of dye over bare, sandy seabeds and underwater sea-grass meadows. We blobbed shallow water and deeper spots. We wanted to learn how all these factors affected transport. Microscopic larvae are such weak swimmers that the currents move them around the same way they moved the dye."

Mimi also had a camera. "Then we took photographs of the dye blobs spreading around the lagoon. By measuring movement and expansion of dye blobs in those photos, we'd be able to do mathematical calculations on computers back home to figure out how the terrain of the lagoon floor affected how larvae would be carried around by the water."

~ Bathing Suit Blunder

Clad in his fluorescent green swim trunks, Zack checks out a colony of sea squirts in a clump of seaweed on Palau.

Mimi got tired of wearing the same sarong day after day, but another wardrobe glitch proved more serious. "That year all the swimwear was hot pink or fluorescent green," Mimi recalls. "When Zack wore his fluorescent green swim trunks and got caught standing in a patch of fluorescent green dye, I couldn't distinguish between his bathing suit and the blob. So I had to throw out some of our experiments."

"We still got enough data to analyze," continues Zack. "But it took an army of undergraduate assistants at Berkeley to digitize the blobs from hundreds of photographs. A computer then calculated the precise patterns and directions in which the blobs moved. We spent years analyzing experiments that took us only about three weeks to do."

But the experiments had their effect on Mimi. Having observed how ocean currents transport larvae, she started to think about looking at the larvae themselves. Mimi got her chance in 1997 when she took a sabbatical at the University of Hawaii.

~ Nosing Around for Noses

When Mimi headed to Hawaii, she planned to study sea-slug noses with marine zoologist Mike Hadfield. "They're not hairy like lobster noses," Mimi explains. "They're smooth posts. I wondered how that kind of nose caught odor molecules. Could they sniff?"

When Mimi learned that Mike was teaching a graduate-level course on marine larvae, she took it to learn more about larvae. Mimi continues, "Mike talked a lot about how chemical cues make certain larvae settle down on the ocean floor and undergo metamorphosis—the process where they change into their adult forms."

Mike explained that the cues are odors given off by organisms on the seafloor. One cue is "perfume" given off by adults of the same species as the larvae. Another is the aroma of food eaten by bottom-dwelling adults.

Does the head *(below, far left)* of the sea slug *Phestilla sibogae* remind you of a snail? If so, it's because the two creatures are related. This slug, about an inch long, eats *Porites* coral.

Mimi hit Mike with a friendly challenge. "Every experiment ever done about cues has taken place in a glass beaker inside a lab," she said. "The results don't say anything about the real world of swirling sea currents and crashing waves. Maybe the ocean mixes up a cue so much that it has no effect on the larvae. Besides, we all know that larvae are crummy swimmers. Even if they smelled a cue, how could they do anything about it? Let's figure this out for ourselves."

Mike accepted the challenge. Instead of working on sea-slug noses, Mimi and Mike collaborated on sea-slug larvae.

The two scientists picked a sea slug known by the scientific name of *Phestilla sibogae* as their "model species." Mike had been working on *Phestilla* for a long time. He knew from beaker tests that *Phestilla* larvae change form when they smell *Porites*—a coral species that is the favorite prey of sea-slug adults. Mimi and Mike now posed two questions: How does the ocean water flowing over a coral reef mix up and dilute the aroma coming from *Porites*? And what happens to sea-slug larvae in that turbulent, wavy flow?

~ *Dropping on Cue*

The one-inch-long *Phestilla* produces up to 6,000 eggs a day. So rearing lots of larvae in the lab was possible. But experimenting with them presented a challenge. Each larva is so tiny—a fraction of a grain of pepper—that scientists measure it in microns, or millionths of a meter.

Mimi looked at the tiny larvae under a microscope. "They look like little Mickey Mouses," laughs Mimi. "They have flaps that resemble big ears. Cilia, or microscopic hairs, run like fringe around the edge of each flap. These cilia beat back and forth to propel the larvae through the water."

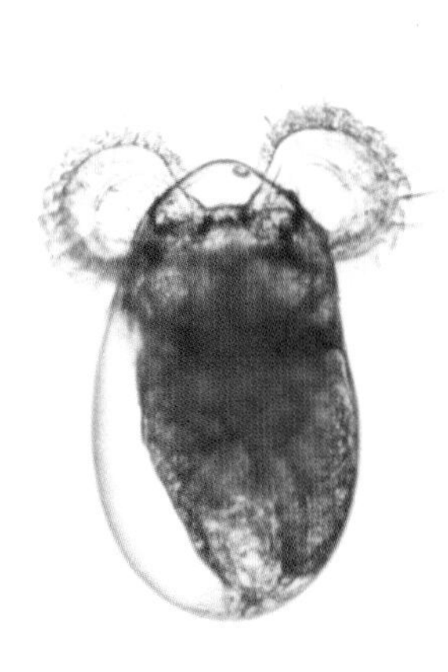

A *Phestilla* larva appears 175 times its normal size in this image taken through a microscope. Twin retractable swimming flaps give the creature the look of a miniature Mickey Mouse.

To figure out if these larvae ran into cue, Mimi and Mike moved from the lab to the ocean. They wanted to find out what happened to the odors given off by *Porites* coral. The pair of scientists placed scientific instruments in and above a coral reef in Kaneohe Bay, Hawaii. Readings from these precise gauges told them that water flows not only over the reef, but through it (a coral reef is porous). "We hypothesized that the water has plenty of time to pick up *Porites* perfume as it flows through the spaces among branches in the reef," explains Mimi.

To test their hypothesis, the two scientists took water samples inside the reef, directly above it, and at varying distances away. Then they returned to the lab, where they exposed larvae to the various samples. A simple reaction told them when a larva had encountered cue-rich water: It dropped to the bottom of the beaker and underwent metamorphosis. "As we guessed," says Mimi, "cue exists within the reef itself and right above it."

~ Stirring Things Up

Knowing where chemical cue was found did not tell Mimi and Mike how sea-slug larvae act when they run into it. Mimi recalled the project she had tackled with engineer Jeff Koseff—the one that revealed "stripes of stink" sniffed by lobsters. "I assumed that larvae swim through similar filaments of coral aroma," Mimi explains. "So I teamed up with Jeff to see what odor filaments look like above a wave-washed coral reef."

Jeff's graduate student, Matt Reidenbach, used skeletons of dead corals that Mike shipped from Hawaii to build a reef inside a big flow tank. He then set water flow to mimic the waves and turbulence that Mimi had measured over real reefs.

Because the scientists didn't use living corals, they needed to re-create cue leaking from the reef. They decided to paint it with a mixture of gelatin and fluorescent dye. "As the gelatin slowly dissolved," Mimi recalls, "it released dye into the water in much the same way that cue comes off of corals. We then used a laser to 'slice' a very skinny sheet from the dye cloud. Not only did that show the dye from the larva's point of view, it let us make a moving map of the stripes of cue a larva runs into as it drifts over a reef."

~ Sink or Swim

Mimi and Mike still didn't know what sea-slug larvae do when they encounter coral cue stripes. So just as Mimi had done when studying copepods, they monitored a larva's movements by putting it on a leash. "Using petroleum jelly," Mimi notes, "we stuck a very fine wire to a larva's shell. We then used a video-and-microscope setup to film the creature."

From previous observations, Mimi's team knew that the Mickey Mouse flaps of free-swimming larvae propel the tiny creatures through the water at a particular speed. So the team moved water in a little flow tank past the leashed larva at the same speed.

Mimi then released stripes of coral cue into the water, capturing the larva's response on video. "Whenever a larva met the odor," Mimi exclaims, "it pulled in its swimming flaps and began to sink like a little stone. When the larva moved into an area without cue, by contrast, its flaps popped out again and the larva started swimming once more. Basically, the larvae are essentially on-off machines: When they smell *Porites,* they sink; when they don't, they swim."

Adult *Phestilla* sea slugs munch on their favorite food, the coral *Porites* *(above)*. A beautiful expanse of *Porites* is shown in the top photo.

The findings surprised Mimi. "I thought the cue got mixed up so much it wouldn't affect where the larvae settled," she admits. "Even if the larvae did run into some cue, I thought, 'What could these puny creatures do about it?' All my ideas were wrong! But that was great. We wound up learning how this on-off mechanism helps larvae bump into a reef."

Mimi and Mike went on to learn that coral cue causes sea-slug larvae to secrete a sticky substance that helps them adhere to the reef. Such studies excited Mimi so much that she's since stuck around Hawaii each summer to study how different types of marine larvae find their homes in different habitats. She's also gone off in other directions, stopping for a few new honors along the way.

In many ways,
Mimi has **created**

her **own** *brand of science.*

12

A Place in Science

On May 1, 2001, Mimi climbed out of bed before sunrise to work at her computer. Around 6 A.M, a ringing telephone jangled her nerves. *Nobody calls this early unless it's an emergency*, she thought as she raced for the phone.

Picking up the receiver, Mimi learned that the National Academy of Sciences (NAS) in Washington, D.C., had just elected her as a member. A friend of Mimi's had run out of the early morning meeting to tell her about the honor.

"I just kept pinching myself," Mimi exclaims. "No sooner had I hung up the phone than it rang again. I got maybe five or six calls. I recognized all the voices, so I knew it was true."

Because of the three-hour time difference between the West Coast and the East, Mimi fielded the calls while many people in Berkeley, California, were still sleeping. One of those people was Mimi's husband. She wanted to wake Zack with the good news, but he had just logged a string of late nights finishing a grant proposal.

"Here's poor Zack, getting his first sleep in days, and I'm tiptoeing around just bursting with this news," Mimi recalls. "In a field like mine, where there's no Nobel Prize, election to the National Academy of Sciences is the biggest honor you can receive. I would have found people at work, but I wanted to tell Zack more than anybody in the world."

What would Einstein think? Mimi *(opposite, far left)* and six other members of her women's group climb into the lap of a huge statue of Albert Einstein outside the National Academy of Sciences. Above, Mimi proudly adds her name to the academy's membership book.

The minutes ticked by. Finally, around 10 A.M., Zack began to stir. As soon as he was semiconscious, Mimi excitedly shared her news. Zack took Mimi out on the deck of their house, which overlooks San Francisco Bay, and the two of them stood watching the morning sun glint off the Golden Gate Bridge.

"I don't remember my exact words," smiles Zack. "But it went something like this: 'Remember that you're good. Even though you may not quite believe it, others do.'"

The morning sun highlights San Francisco and the Golden Gate Bridge in this panoramic view from the deck of Mimi's house in Berkeley.

~ *Making a Mark*

Despite Zack's words, Mimi felt so overwhelmed by the NAS honor that some of her old doubts came creeping back. "It was like winning the MacArthur Fellowship in 1990," recalls Mimi. "Part of me said, 'There must be some mistake.'"

People soon persuaded Mimi otherwise. "My sister always surprises me when she doubts what she's achieved," laughs her brother, Bob. "The NAS selects about 70 people a year—scientists who've done all kinds of incredible things. Then, in 2001, my sister was among them. Yeah, she's special, and I told her so."

Mimi's mentors, colleagues, and friends felt the same way. "I went up and down our department at Duke University announcing the good news," says Mimi's Ph.D. advisor, Steve Wainwright. "Only one other Duke zoology student—a male—had ever collected the NAS award. I couldn't have been happier."

The NAS, created by President Abraham Lincoln in 1863, advises the federal government on scientific and technical matters.

Membership in the group is based on a lifetime of accomplishment. As of 2005 only about 2,000 people belonged to the organization.

Bob Paine, a longtime NAS member, supported Mimi's nomination. "You have to be productive to get into the NAS," he says, "but two other things rank higher: the novelty of a person's work and the impact of that work. When NAS members review a nominee, they ask, 'Has this individual made a difference scientifically?' Mimi surely has."

~ *Butterflies Among the Beetles*

Because the NAS induction ceremony wouldn't take place until 2002, Mimi had time to plan for the event. She invited members of her women's problem-solving group, four of whom now belong to the NAS. She also asked her family and other close friends to attend.

Even with this support, Mimi got jittery as the induction neared. She wondered, *What on earth should I wear?* She did some "girl talking" with Inez Fung, a Berkeley climate expert who had won the NAS award the same year. Inez planned to wear her mother's brilliant green Chinese wedding dress. Mimi picked a fancy black dress and the flowing, hand-painted silk jacket she had worn at her own wedding.

It's official! Mimi accepts the congratulations of Bruce Alberts, President of the National Academy of Sciences, upon being inducted into the academy in 2002.

When the big day arrived, the two women stood out in a sea of dark-suited men. Laughs Mimi's friend Suzanne McKee, "They looked like butterflies among a crowd of beetles."

The inductees and their guests took their seats in a huge auditorium. Then, one by one, the scientists walked across a broad stage to sign an enormous book. "That book," explains Mimi, "contains the names of every person ever elected to the academy. It overflows with the signatures of famous scientists from the past."

When Mimi's turn came, she proudly added her name to the book. Only a small percentage of the signatures in it belong to women, reflecting their long struggle to break into the sciences. Yet those who watched Mimi walk across the stage had gathered to celebrate the arrival of one more butterfly among the beetles—a woman who has shown that the words "female" and "scientist" belong together.

~ *Into the Limelight*

Mimi's brother, Bob, and his wife, Mary Koehl—the two people who had supported Mimi through some of the most difficult years of her life—hosted a party for Mimi. As a surprise guest they invited Mimi's sixth-grade teacher, Harriett Whittaker. In keeping with the family tradition of practical jokes, Bob urged her to tell some embarrassing "kid story" about Mimi. "I don't have a single bad thing to say about your sister," Harriett gasped, "and you know it, Bob Koehl!"

Mimi's brother, Bob, and his wife, Mary, have played key roles in Mimi's life. "Mary has always been there for me in times of trouble," recalls Mimi. "Her wisdom helped me sort out the pain of troubled teenage years, and it helped me cope with losing both my parents at a young age."

The day after her NAS induction, Mimi attended another party where people kept congratulating her. It turned out they were referring to her newly announced election to the American Academy of Arts and Sciences, founded by Harvard College graduates in 1779.

"I had no idea this would happen," exclaims Mimi. "Inductees that year included Senator Edward Kennedy, African novelist Chinua Achebe, and actress Anjelica Huston, not to mention a host of scientists. It would be like the MacArthur Fellowship: I'd get to see my heroes."

But Mimi wasn't done collecting awards in 2002. The American Society of Biomechanics awarded her its Giovanni Borelli Award.

(Borelli studied physics and human movement during the time of Galileo.) "The society gives the award each year to somebody who has greatly advanced the field of biomechanics," Steve Wainwright explains. "It represents genuine appreciation by your peers."

~ *Encore! Encore!*

Put yourself in Mimi's shoes after the whirlwind year of 2002. What would you do as an encore? When asked that question, Mimi smiles, "I kept doing what I love—looking at how things work in the natural world and teaching students how to do it too."

Mimi has continued to work with lobster noses, larvae, and other research begun earlier in her career. But she still does new projects for "fun." In 2004, for example, she got excited when scientists uncovered fossils of feathered dinosaurs more than 120 million years old. She wondered, *How did those feathers affect the movement of those animals through the air?*

> Only a small percentage of the signatures in it belong to women, reflecting their long struggle to break into the sciences.

Mimi used this question to tickle the curiosity of Karen Yang, an undergraduate student who applied to work in Mimi's lab. Mimi invited Karen to take a feathered dinosaur on a test flight. Karen, a bioengineering major, jumped at the chance to jump into the past.

The fossil that captured Mimi's imagination is a species called *Microraptor gui*, or *M. gui*. Not only did this dinosaur sport feathers on its arms, as do modern birds, but it also had feathers on its legs. Still more feathers fringed its long dinosaur tail. "I love *M. gui* because it was such a beautiful beast," exclaims Mimi.

~ *Flexible Feathered Mummy*

Based on fossils of *M. gui*, Mimi and Karen built a model of a feathered dinosaur to see what it could and could not do mechanically. They fashioned it out of Styrofoam, paper tape,

The fossil *(above)* is the feathered dinosaur, *Microraptor gui.* This 30-inch-long stone imprint served as a blueprint for the model built and tested by Mimi and bioengineer Karen Yang, who is shown at right using an anemometer. This device measures the speed of the air moving across the model's feathered arms and legs as it "glides" in the wind tunnel.

and bird feathers. "Basically, it's a flexible feathered mummy," Mimi laughs. "We can twist it into different shapes and add or remove feathers for our various experiments."

To run these experiments, Mimi and Karen had read geology papers to discover what Earth's atmosphere and climate were like 120 million years ago. Air composition and temperature can affect how well airborne animals glide, parachute, or steer. "It's like being a detective looking for clues," Mimi declares, "only we were looking for clues to help us see the world through the eyes of *M. gui.*"

Mimi and Karen took their feathered model on its first test flight in 2004. But as of 2005, they didn't have enough data to answer their questions about *M. gui*. Looking to the future Mimi declares, "Science is an ongoing process. Sometimes we get frustrated with the slow pace of research. But it's the frustration of not knowing—of unanswered questions—that keeps us doing experiments and learning something new, something unexpected."

~ So What?

Recall the crystal ball technique that Steve Wainwright used to teach students like Mimi. You might ask: So what? Suppose I had a crystal ball and could see all the answers about how nature's living machines work. What would I know?

Well, for one thing, you'd learn the link between engineering and biology. You'd realize, for example, why organisms with certain body designs thrive in particular habitats better than organisms with different designs. You'd get an idea of how human beings can change a habitat so a living machine doesn't work anymore.

What has Mimi invented? She's devised some creative techniques, but she isn't concerned with practical inventions for human use. "Basic researchers like me," Mimi points out, "hope to discover general principles about how the natural world works: How do stars, mountains, or babies form? What drives chemical reactions? How do animals swim, fly, and smell?"

Applied researchers put these principles to work. These scientists solve practical problems by applying the discoveries turned up by basic researchers. Using what Mimi has unveiled about the operation of lobster noses, for example, engineers may one day be able to build mechanical robot noses.

Remember that, in many ways, Mimi has created her own brand of science. She went beyond existing scientific fields to look at the physics behind natural shapes, forms, and movements that captured her imagination.

When talking to kids, Mimi offers this advice: "Every person has some gift or talent that can help unlock new answers about the world in which we live—and the organisms we share it with. So don't be afraid of science, but do tackle it the way you do best. It's a big discipline, but you can still find your own special spot in science just like I did."

Timeline of Mimi Koehl's Life

1948 Mimi Alma Ruth Koehl is born on October 1 in Washington, D.C.

1953 She attends public school for the next 13 years in Silver Spring, Maryland.

1956 Mimi's older brother, Bob, leaves for college.

1966 Mimi graduates 18th—the same rank as her brother—in a high school class of more than 1,000 students. She goes to Gettysburg College in Pennsylvania.

1970 She receives the Columbia Teachers College Book Prize for her work in elementary education. After switching her major from art to science, Mimi graduates with honors from Gettysburg College with a bachelor's degree in biology. She begins her graduate studies in zoology at Duke University.

1971 Steve Wainwright returns from sabbatical and becomes Mimi's mentor, inspiring her to enter the field of biomechanics.

1972 Mimi meets ecologist Bob Paine and begins her field studies of wave forces on organisms on the shore.

1976 Duke awards Mimi a Ph.D. based on her research on sea anemones.

1977 Peritonitis and appendicitis nearly claim Mimi's life while she is a postdoc at Friday Harbor Laboratories. Mimi moves to England for a year of postdoctoral study at the University of York.

1978 Mimi returns from England to become an assistant professor at Brown University in Providence, Rhode Island.

1979 Mimi joins the faculty at the University of California at Berkeley. She appears in a *NOVA* television program on biomechanics.

1981 Mimi starts meeting with a group of other women to discuss career-related challenges. They still meet today.

1982 She is put up for tenure three years early.

1983 Mimi wins the Presidential Young Investigator Award for her promising career in science.

1985 Gettysburg College awards Mimi its Young Alumni Achievement Award.

1986 Mimi returns to England as a visiting scholar at the Center for Mathematical Biology at Oxford University.

1988 Mimi wins a John Simon Guggenheim Memorial Foundation Fellowship.

1990 The MacArthur Foundation picks Mimi to receive a fellowship to pursue creative science.

1993 Mimi weds oceanographer Zack Powell.

2001 The National Academy of Sciences elects Mimi as a member.

2002 The American Academy of Arts and Sciences elects Mimi as a member.

2002 The American Society of Biomechanics gives Mimi its Borelli Award for "outstanding career accomplishment" and "exemplary contributions to the field of biomechanics."

2005 Mimi continues researching how lobsters smell and how marine larvae find homes. She is also examining new questions, including how corkscrew-shaped bacteria get twisted up and how feathers might have affected the flight of dinosaurs.

About the Author

Deborah Amel Parks—author, journalist, and world traveler—got interested in science during the sixth grade. A teacher taught her class outdoors so students could see the adventures taking place around them. She's since backpacked all over the world and climbed mountains on every continent except Antarctica. Deborah has written seven children's books and numerous magazine and newspaper articles. She's also been a contributor on many major educational projects. Deborah lives in New York's Hudson Valley with her husband, Richard Parks, and their many animals, including seven cocker spaniels.

Glossary

This book traces a new science called biomechanics. You can probably guess what this means by looking at two word parts: *bio* and *mechanics*. The first word part comes from the Greek word *bios*, meaning "life." The second word part means the study of the physics of force and motion. Most biomechanists study how human machines work. But some scientists in this field study how other living machines work—microscopic sea creatures, kelp, lobsters, flying frogs, and even extinct animals.

This science is a bit different than your normal biology class. So you may find some unfamiliar terms. Take them apart or use context clues to figure them out. Here are some terms to get you started. For more information about them, consult your dictionary.

aerodynamics: the study of the motion of fluid gases, like air, and the forces that act on bodies exposed to them; engineering and physics principles used to design airplanes, for example.

biomechanics: the use of principles of engineering and the laws of physics to study the movements and designs of living organisms

boundary layer: a layer of water (or air) close to the ground or any other surface that moves more slowly than water (or air) rushing overhead

copepods: tiny, planktonic, shrimp-like animals found in oceans and lakes. They are about the size of a sesame seed.

drag: the resistance of a fluid—air or water—to an object in motion; the force a moving fluid exerts on a body tending to push it downstream.

flow tank: a long aquarium-like tank through which water flows. The speed and turbulence of the water in the tank can be controlled and changed.

fluid: a liquid or gas that flows easily and takes the shape of its container

fluid dynamics: the study of how liquids and gases move

hydrostatic skeleton: the mechanical support system of soft-bodied animals, like worms and sea anemones. A hydrostatic skeleton is made up of a muscular body wall surrounding a fluid-filled cavity, like a water balloon.

intertidal zone: the area on a shore between the upper and lower tide marks

invertebrate: an animal without a spinal column. This includes animals that are not fish, amphibians, reptiles, birds, or mammals.

kelp: certain types of large, brown seaweeds

larvae: creatures that at birth or hatching look very different from their parents. Frog tadpoles are an example.

mesogloea: the tissue surrounding the water-filled gut of sea anemones

oceanography: the study of the ocean, its currents, its chemistry, its geology, and the things that live in it

plankton: tiny plants and animals that live in the water in oceans and lakes

rogue wave: a large wave formed when faraway storms push smaller waves together

settlement cue: a chemical odor given off by bottom-dwelling marine life. A cue can tell larvae where to settle or can induce them to metamorphose into bottom-dwelling juveniles.

solid: a material that has a definite shape and does not flow

spicules: needlelike skeletal structures that appear in a variety of animals like sponges and soft corals. Spicules are made of silica (glass) or calcium carbonate (chalk).

stipes: stems of seaweeds

wind tunnel: a chamber through which air is blown at set speeds over a test object, allowing scientists to gather data on aerodynamics

zoology: the branch of biology concerned with the animal kingdom

Metric Conversion Chart

When you know:	Multiply by:	To convert to:
Inches	2.54	Centimeters
Feet	0.30	Meters
Miles	1.61	Kilometers
Acres	0.40	Hectares
Pounds	0.45	Kilograms
Centimeters	0.39	Inches
Meters	3.28	Feet
Kilometers	0.62	Miles
Hectares	2.47	Acres
Kilograms	2.20	Pounds

Further Resources

Women's Adventures in Science on the Web

Now that you've met Mimi Koehl and learned all about her work, are you still wondering what it would be like to be a biomechanist? How about a robot designer, a wildlife biologist, or an astronomer? It's easy find out. Just visit the *Women's Adventures in Science* Web site at **www.iWASwondering.org.** There you can live your own exciting science adventure. Play games, enjoy comics, and practice being a scientist. While you're having fun, you'll also get to meet exciting women scientists who are changing our world.

BOOKS

Berger, Melvin, Gilda Berger, and John Rice. *What Makes an Ocean Wave? Questions and Answers About Oceans and Ocean Life.* New York: Scholastic Inc., 2001. Enter Mimi's underwater world and learn the answers to 60 important questions about oceans. Besides telling you how oceans wave, you'll uncover all kinds of interesting facts about tides and undersea animals.

Lafferty, Peter. *Eyewitness: Force and Motion.* New York: Dorling Kindersley, 1999. Does the word "physics" scare you? If so, you might want to check out this book that takes the fear out of this branch of science. You'll be surprised by how physics affects our everyday lives.

Royston, Angela. *Moving.* My Amazing Body Series. Chicago: Raintree, 2004. Want to know how your own living machine—the muscles and bones that make up your body—gets into motion? If so, check out this handy, 32-page guide to the mechanics behind our unique engineering and how it allows us to function.

WEB SITES

Friday Harbor Laboratories: http://depts.washington.edu/fhl/
Travel across the United States to visit the marine labs where Mimi built her first flow tank and plunged deeper in the mysteries of how calm-water sea anemones work. Skim through some of the courses offered. Did you ever think about going to school at a place like this?

The Koehl Lab: http://ib.berkeley.edu/labs/koehl/
Peak inside Mimi's laboratory at Berkeley. Learn about the research that she and her students are doing. See pictures of folks in action in the lab. Find links to other scientists doing biomechanics, marine biology, or biophysics.

Nudibranches: http://oceanlink.island.net/oinfo/nudibranch/nudibranch.html
Mimi could spend a lifetime studying sea-slug larvae and still not get to all of the species—some 3,000. Learn how these beautiful, diverse sea animals eat, develop, reproduce, and defend themselves. Then find out where to see them for yourself.

Women oceanographers: www.womenoceanographers.org/
Meet and get to know the work of inspiring marine scientists who are contributing to our understanding and appreciation of our oceans and the life in them.

Woods Hole Oceanographic Institution: www.whoi.edu
Take a tour of Earth's oceans by visiting the world's largest independent oceanographic institute at Woods Hole in Massachusetts. It's the place to go if you're interested in following a career in marine biology or oceanography.

SELECTED BIBLIOGRAPHY

In addition to interviews with Mimi Koehl, her family, friends, and colleagues, the author did extensive reading and research to write this book. Here are some of the sources she consulted.

Gordon, J. E. *Structures: Or, Why Things Don't Fall Down*. 2nd ed. New York: Da Capo Press, 2003.

Koehl, Mimi A. R. *Wave-Swept Shore: The Rigors of Life on a Rocky Coast*. Berkeley, CA: University of California Press, 2006.

Shapiro, Ascher H. *Shape and Flow: The Fluid Dynamics of Drag*. Garden City, NY: Anchor Books, 1961.

Vogel, Steve, and Kathryn K. Davis (Illustrator). *Cats' Paws and Catapults: Mechanical Worlds of Nature and People*. New York: W. W. Norton & Company, 2000.

Vogel, Steve. *Comparative Biomechanics: Life's Physical World*. Princeton, NJ: Princeton University Press, 2003.

Vogel, Steve. *Life's Devices: The Physical World of Animals and Plants*. Princeton, NJ: Princeton University Press, 1988.

Vogel, Steve, and Annette deFerrari (Illustrator). *Prime Mover: A Natural History of Muscle*. New York: W. W. Norton & Company, 2003.

Wainwright, Stephen A. *Axis and Circumference: The Cylindrical Shape of Plants and Animals*. Cambridge, MA: Harvard University Press, 1988.

Wainwright, Stephen A., et al. *Mechanical Design in Organisms*. Princeton, NJ: Princeton University Press, 1982.

INDEX

Women's Adventures in Science Advisory Board

A group of outstanding individuals volunteered their time and expertise to help the Joseph Henry Press conceptualize and develop the *Women's Adventures in Science* series. We thank the following people for their generous contribution of time and talent:

Maxine Singer, Advisory Board Chair: Biochemist and former President of the Carnegie Institution of Washington

Sara Lee Schupf: Namesake for the Sara Lee Corporation and dedicated advocate of women in science

Bruce Alberts: Cell and molecular biologist, President of the National Academy of Sciences and Chair of the National Research Council, 1993–2005

May Berenbaum: Entomologist and Head of the Department of Entomology at the University of Illinois, Urbana-Champaign

Rita R. Colwell: Microbiologist, Distinguished University Professor at the University of Maryland and Johns Hopkins University, and former Director of the National Science Foundation

Krishna Foster: Assistant Professor in the Department of Chemistry and Biochemistry at California State University, Los Angeles

Alan J. Friedman: Physicist and Director of the New York Hall of Science

Toby Horn: Biologist and Co-director of the Carnegie Academy for Science Education at the Carnegie Institution of Washington

Shirley Ann Jackson: Physicist and President of Rensselaer Polytechnic Institute

Jane Butler Kahle: Biologist and Condit Professor of Science Education at Miami University, Oxford, Ohio

Barb Langridge: Howard County Library Children's Specialist and WBAL-TV children's book critic

Jane Lubchenco: Professor of Marine Biology and Zoology at Oregon State University and President of the International Council for Science

Prema Mathai-Davis: Former CEO of the YWCA of the U.S.A.

Marcia McNutt: Geophysicist and CEO of Monterey Bay Aquarium Research Institute

Pat Scales: Director of Library Services at the South Carolina Governor's School for the Arts and Humanities

Susan Solomon: Atmospheric chemist and Senior Scientist at the Aeronomy Laboratory of the National Oceanic and Atmospheric Administration

Shirley Tilghman: Molecular biologist and President of Princeton University

Gerry Wheeler: Physicist and Executive Director of the National Science Teachers Association

Library Advisory Board

A number of school and public librarians from across the United States kindly reviewed sample designs and text, answered queries about the format of the books, and offered expert advice throughout the book development process. The Joseph Henry Press thanks the following people for their help:

Barry M. Bishop
Director of Library Information Services
Spring Branch Independent School District
Houston, Texas

Danita Eastman
Children's Book Evaluator
County of Los Angeles Public Library
Downey, California

Martha Edmundson
Library Services Coordinator
Denton Public Library
Denton, Texas

Darcy Fair
Children's Services Manager
Yardley-Makefield Branch
Bucks County Free Library
Yardley, Pennsylvania

Kathleen Hanley
School Media Specialist
Commack Road Elementary
Islip, New York

Amy Louttit Johnson
Library Program Specialist
State Library and Archives of Florida
Tallahassee, Florida

Mary Stanton
Juvenile Specialist
Office of Material Selection
Houston Public Library
Houston, Texas

Brenda G. Toole
Supervisor, Instructional Media Services
Panama City, Florida

Student Advisory Board

The Joseph Henry Press thanks students at the following schools and organizations for their help in critiquing and evaluating the concept for the book series. Their feedback about the design and storytelling was immensely influential in the development of this project.

The Agnes Irwin School, Rosemont, Pennsylvania
La Colina Junior High School, Santa Barbara, California
The Hockaday School, Dallas, Texas
Girl Scouts of Central Maryland, Junior Girl Scout Troop #545
Girl Scouts of Central Maryland, Junior Girl Scout Troop #212

Illustration Credits:

Except as noted, all photos courtesy Mimi Koehl

Cover, viii © 2004 Lise Dumont; **1** Mike Hadfield; **2** Courtesy Bob Paine; **7** *(t)* Museum of the Rockies; *(b)* Steve Moore; **12** *(t)*, **16** Mimi Koehl; **18** *(t)* Terry Wild Studio; **20, 21** Mimi Koehl; **24** D. Weisman/© Woods Hole Oceanographic Institution; **26** Les Todd/Duke University Photography; **27** Ken Sebens; **31** Courtesy Steve Wainwright; **33** © 2000 PhotoDisc; **34** Mimi Koehl; **35** NOAA; **36** Robert Steelquist, Olympic Coast National Marine Sanctuary; **37** Courtesy Bob Paine; **38** *(t)* Robert Steelquist, Olympic Coast National Marine Sanctuary; *(b)* Ken Sebens; **39, 40, 43-46, 47** *(b)* Mimi Koehl; *(t)* Steven Fisher; **50** A. Richard Palmer; **51** Mimi Koehl; **52** Ken Sebens; **53** © 1992 PhotoDisc; **55** *(t)* © Friday Harbor Laboratories; *(b)* Megumi Strathmann; **58** Skycell Ltd.; **59** Courtesy John D. Currey; **60** *(t)* Jim Greenfield/imagequestmarine.com; *(b)* Mimi Koehl; **66** Courtesy Ellen Daniell; **68** *(t)* Mimi Koehl and G. Paffenhofer; *(b)* Mimi Koehl; **70** *(t)* Mimi Koehl; *(b)* Zack Powell; **72** © 2000 PhotoDisc; **74, 75** Mimi Koehl; **76** Courtesy Joel Kingsolver; **78** Mary Koehl; **80** Mike McCay; **82** Ellen Daniell; **83** Bob Koehl; **87** Jim Higgins; **88** Mimi Koehl and Jeff Koseff; **90** Mike Hadfield; **91-93** Mimi Koehl; **94, 95, 97** *(t)* Mike Hadfield; *(b)* Robin B. Kinnel, Professor of Chemistry, Hamilton College; *(inset)* Mike Hadfield; **98** Steven Gross; **99** James H. Pickerell; **100** Mimi Koehl; **101** James H. Pickerell; **104** *(t)* Mick Ellison; *(b)* Courtesy Karen Yang

Maps: © 2004 Map Resources

Illustrations: Max-Karl Winkler

The border image used throughout the book is of a coral reef and is from © 1999 PhotoDisc

JHP Executive Editor: Stephen Mautner

Series Managing Editor: Terrell D. Smith

Designer: Francesca Moghari

Illustration research: Christine Hauser

Special contributors: Kristine M. Calo, Meredith DeSousa, Olaf Ellers, Allan Fallow, John Gosline, Sally Groom, Mary Kalamaras, Dorothy Lewis, Rachel Ann Merz, John Quackenbush

Graphic design assistance: Michael Dudzik